MANUFACTURING TECHNOLOGY TRANSFER

A Japanese Monozukuri View of Needs and Strategies

MANUFACTURING TECHNOLOGY TRANSFER

A Japanese Monozukuri View of Needs and Strategies

Yasuo Yamane • Thomas Childs

CRC Press is an imprint of the
Taylor & Francis Group, an **informa** business

A PRODUCTIVITY PRESS BOOK

CRC Press
Taylor & Francis Group
6000 Broken Sound Parkway NW, Suite 300
Boca Raton, FL 33487-2742

Printed on acid-free paper
Version Date: 20130204

International Standard Book Number-13: 978-1-4665-6763-4 (Hardback)

Library of Congress Cataloging-in-Publication Data

Yamane, Yasuo.
Manufacturing technology transfer : a Japanese monozukuri view of needs and strategies / Yasuo Yamane and Thomas Childs.
p. cm.
Includes bibliographical references and index.
ISBN 978-1-4665-6763-4
1. Production management--Japan. 2. Technology transfer--Japan. 3. Production engineering--Japan. 4. Manufacturing--Japan. I. Childs, Thomas (Thomas H. C.) II. Title.

TS155.Y26 2013
658.5--dc23 2012036474

Visit the Taylor & Francis Web site at
http://www.taylorandfrancis.com

and the CRC Press Web site at
http://www.crcpress.com

Contents

Preface

This book is written for the following three readerships:

- People who are interested in how to transfer manufacturing technology to next generations of manufacturing industry engineers
- People who are interested in the overseas operation of Japanese manufacturing industries
- Students who are studying manufacturing technology management courses, for example, at the master's level, or who are interested in employment in manufacturing industry

It is written from a Japanese manufacturing management point of view. Its particular focus is on manufacturing technology transfer to South East Asia. This region, along with, for example, India, is growing as a major manufacturing consumer as well as a manufacturing producer. Manufacturing technology transfer also requires the transfer of skills to use the technology. The contents of this book concerned with skills transfer have a much wider relevance than only to South East Asia.

This book is based on a book originally published in the Japanese language by one of us (Y.Y.) with colleagues mainly from Hiroshima University. Its title, in translation, is *Manufacturing Industry's Overseas Operation from the Point of View of Technology and Skill Transfer* (Tokyo: Nikkan Kogyo Shimbu, 2008). It was planned that a rough translation of that would be produced by a Japanese speaker. This would then be improved by a native English speaker who also knew its subject area. The other of us (T.H.C.C.) is that person. Finally, the resulting English text would be checked with the original Japanese authors and corrections would be made in cases where the meaning was lost. This is what happened. But the final stage—checking with the Japanese authors—resulted in some cases in major revisions to the original text. Some parts were lengthened and others shortened. Content was moved between chapters. The order of chapters was changed. New material was added. Corrections were made. Hence, this book may be thought of as a second edition with altered content as well as language compared with the original Japanese text.

For the first readership (people interested in the development of future manufacturing engineers for industry), the book proposes a general way of describing the transfer of technology and skills from teacher to learner. A general way is necessary because a wide range of technologies and skills are needed in manufacturing industry. All departments of manufacturing industry need and carry out staff training. Human resource development in industry needs not only methods optimized for individual circumstances but also general methods that enable comparisons and quantitative evaluations to be made between and across departments. Without such generalized methods it is difficult to set up master plans that are essential to large business activities.

The general way is by means of a learning curve. The book develops a general description of obtaining technology or skill in manufacturing in terms of a response (the learning curve) to a teaching input. The learning curve can be quantified in the same way that the response curve to a unit step input to an automatic control system is quantified. It discusses how such generalized methods can benefit both workers and employers.

For the first readership also, the book shows an example of how the learning time for skilled work may be shortened by using information technology (IT). This is a research interest of the book's authors. Manufacturing technologies, monozukuri technologies in Japanese, are technologies through which artifacts are created. Therefore, those who engage in manufacturing need creative abilities as well as the skills to realize the artifacts. Machines and computers developed in recent years support creative work but do not carry out the work instead of people. Computer-aided design (CAD) is a very strong tool for design, but CAD does not create the new ideas and their achievement. Only the substantial abilities of humans can do that.

The required high-level skills can take a long period of training for their development. In other fields IT has proved to be a powerful tool to shorten training times. An example is the use of flight simulators to develop flying skills. The book introduces an example of a simulator to speed up learning manufacturing skills, namely, a scraping simulator. Scraping is a surface finishing technology or skill used in precision and ultra-precision manufacturing industry such as the precision machine tool industry and the ultra-precision machine industry. Today's precision machines could not be realized without scraping skills.

For the second readership (people interested in the international activities of Japanese manufacturing companies), the book contains the results of research survey work carried out at Hiroshima University between 2005 and 2007, under the title *Overseas Operation of Japanese Manufacturing Industries*. It was supported by the Ministry of Economy, Trade and Industry (METI) in Japan. The investigation was carried out by a group made up of members of Hiroshima's Graduate School of Engineering and Center for Collaborative Research and Cooperation. The members of the group and the authors of the original Japanese book's chapters are almost the same as each other.

At that time, the Japanese manufacturing industry faced many problems, such as stagnation of domestic industry caused by overseas deployment, pirating of technology, recall of mass production items such as car battery cells, loss of continuity in technology and skills caused by the retirement of the baby boomer population that was born around 1947–1949, and increase in job turnover by a younger generation of workers. While many problems of overseas operation were identified by the research, the biggest and most urgent issue was found to be developing and retaining human resources to make high-quality products. The issue is important not only for overseas operation but domestically too. The book describes the activities, results, and discussions from that research work.

For the third readership (manufacturing engineering students), the chapters for the first and second groups, specifically the chapters on training means and needs and on overseas survey results, are added to by other chapters. These describe the breadth of manufacturing engineering activities and how manufacturing management styles, priorities, and decision taking depend on a company's position in the marketplace. They also describe the role of legal contracts in supporting successful manufacturing activities across national borders. Case studies illustrate points made in other chapters.

The whole book has been written as a textbook to support a series of lectures on technology transfer within a master's course on management of technology (MOT). Other lecture series are "Venture Business," "Technology Strategy," "Intellectual Property and Finance," and "Innovation Management." "Innovation Management" and "Technology Transfer" are taught not only in Japanese but also in English. The English class is open to international students. Most of the master's course students in engineering take positions in manufacturing industry. Many of them then have a role in overseas operations. It benefits them to learn what technology and skill abilities are required in manufacturing industries, how long it takes to gain these, and to learn about special issues, particularly human issues, arising from operating across national boundaries. These issues include legal, cultural, and infrastructure issues.

While, for the third group, the book is written as a textbook of technology transfer for master's course students, the contents of the book have also been taught in a series of seminars for non-Japanese staff working in overseas bases of Japanese manufacturing industries. It has been well received by them because it helps them to understand the Japanese view of manufacturing, that is monozukuri, technologies.

The expected three readerships described above are different from each other on the surface. But the underlying basic issues of what are the needs of and how to raise the level of the human resources in manufacturing industries are common to all.

In detail, the book consists of 11 chapters.

Chapter 1 introduces the engineering activities of manufacturing industry, for example design, production engineering, machining, and assembly. It traces changes over time, starting with the Industrial Revolution, that have led to increased effectiveness of these activities. These are not only technological changes, for example development of machines of increased accuracy. They are also conceptual changes leading to simplification of tasks, for example simplification of assembly by defining compatible tolerance ranges on part sizes. Compatible tolerances allow the parts to be fitted together easily. Manufacturing industry sectors, for example aerospace, automotive, and electronic device sectors, are classified according to the number of parts that are combined in a product and the accuracy required of them. The abilities that engineers need to work in those industries are outlined. These cover both an ability to understand principles (know-why) and craft skills from learning on the job (know-how). Both types of ability must be taught.

Chapter 2 develops the theme that all teaching and learning processes can be described within a common framework that can be visualized in terms of an S-shaped learning curve. Learning has an initial stage of slow progress (taking some dead time, L), a middle stage of rapid progress (characterized by a time constant, T), and a final stage again of slow progress. These times, L and T, have their equivalents in responses of automatically controlled systems to step inputs. Their magnitudes depend on the difficulty of the activity being learned, the abilities of both the teachers and learners, and on teaching methods too. Skill levels of personnel can be measured in terms of where they are on their learning curves. The resulting data can help workers in evaluating their personal development. It can also help companies in assessing the quality of their workforce and, for example, quantitatively assessing the lost skills costs of staff turnover. The chapter includes typical L and T times for manufacturing industry engineering tasks, obtained from a survey of machine tool builders in Japan. They demonstrate the long times (many years) needed fully to develop high levels of engineering skill.

Chapter 3 develops the idea introduced in Chapter 1 of differences between know-why (explicit knowledge) and know-how (implicit knowledge). It argues that although the relative importance of implicit to explicit knowledge has reduced from ancient times to the present more scientific age, both are still important. How both may be taught and by whom, and how learning times may be reduced, for example by automation or de-skilling (i.e., making tasks easier and hence easier to learn), are general topics of the chapter. The idea of workforce skill level as a valuable company asset, also introduced in Chapter 1, is also considered further. Threats to and how to safeguard that asset at the same time as needing to transfer skills to others, for example in subsidiary or supplier companies, are also topics within this chapter.

Chapter 4 is particularly concerned with how information technology can be used to shorten learning times by off-the-job training, in the same way that flight simulators are used to teach flying skills to pilots. It describes one particular job simulator, namely, a scraping simulator. Scraping is one of the important skills needed for making precision surfaces. By traditional, and still current, methods it can take 10–15 years to become expert at scraping. This particular example acts to highlight some of the challenges within a world-class manufacturing industry that are invisible to those who are not directly involved.

Chapter 5 may be thought of as a continuation of the general introduction of Chapter 1. Whereas Chapter 1 concentrates on the range of engineering technology activities that make up manufacturing, Chapter 5 concentrates on the range of production management activities that support these. These activities include management of product specification and quality, management of production in the quantities and at the times required, and management of the price that can be charged. From these arise classifications of production systems and changes of importance of different management functions at different stages of a product's life cycle. A company's technology strategy will also change through the product's life cycle. All these affect company decisions on overseas expansion. What are the appropriate technologies and skills to be transferred in different circumstances of expansion is a particular topic of the chapter. Problems that arise from overseas' perceptions of transferring companies' motives are taken up in Chapter 8.

Chapter 6 is a factual account of Japanese companies' expansion overseas over the 20 years starting from about 1990. It is based on official Japanese records and the research surveys already mentioned at the start of this preface. It describes the scale of Japanese companies' overseas expansions, with overseas subsidiary companies numbering thousands. It also describes the changing reasons for expansion overseas, from cost reduction at the start of the period to a more global strategy of increasing market share today; and the changes in what technologies are being transferred that accompany those changed reasons for expansion.

Chapter 7 describes how technology transfer overseas should be supported by legal (contractual) agreements between the parties concerned, clearly defining responsibilities and what should be done in the case that problems arise during the transfer. The problems can be both foreseeable and unforeseeable. It is written from a strong personal experience, defining the purpose of agreements, the structure or framework of agreements, and the meaning of legal terms that those involved in agreements need to understand.

Chapter 8 is in one sense a continuation of Chapter 6. Whereas Chapter 6 concentrates on the scale of, reasons for, and content of Japanese manufacturing technology and skills transfers overseas, Chapter 8 is concerned with the views and experiences of the people involved in the transfers. The views are both those of the Japanese carrying out the transfers and the people overseas receiving the transferred technology and skill. Cultural, expectation, and motivation differences between the transferring and receiving participants emerge as some of the factors that can cause tensions between the parties. How these and other human issue problems might be overcome, for example by individual, company, educational, and local or national government actions, to achieve benefits for all, is a theme of the chapter.

Chapter 9 takes up several issues concerning overseas development in manufacturing that can be related to the learning curve introduced in Chapter 2. In overseas manufacturing, it is necessary to secure adequate human resources in the workplace where manufacturing problems have to be solved. There may not be adequate skills to start with. Questions such as how long it takes for basic training of workers, how long it takes for a fully qualified worker to be trained, and what would be the effects of a high turnover rate of staff become important. Decisions on where to set up overseas operations depend on answers to such human resources questions, as well as on labor and materials costs. How the learning curve can be coupled with cost issues to create a quantitative decision-making method on where to locate overseas is the subject of this chapter. It is illustrated by a case study of industrial machinery manufacture.

Chapter 10 presents case studies of Japanese ship building technology and skills transfer overseas. Ship building is an industry where the product has many parts and only a small number of the same product is built at any one time. Know-how is particularly important and problems of technology transfer stem from that. The chapter contains examples of what are considered to be successful and unsuccessful transfer attempts. It considers what were the differences that led to those different outcomes.

Chapter 11 contains another case study, of a Japanese food machinery manufacturer's overseas expansion. It focuses on decision making on where to expand overseas, based on methods described in Chapter 9, and on the long time (measured in years) for a new overseas activity to become established.

We wish to repeat and acknowledge that the original Japanese text is a multiauthor work. This English edition would not have been possible without the full support and collaboration of the original authors. They are:

Chapters 1–3: Professor Yasuo Yamane (Hiroshima University)
Chapter 4: Associate Professor Tadanori Sugino (Oshima National College of Maritime Technology)
Chapter 5: Professor Katsuhiko Takahashi (Hiroshima University)
Associate Professor Katsumi Morikawa (Hiroshima University)
Chapter 6: Professor Tadahiko Takata (Hiroshima University, Former President, Teijin Cord (Thailand) Co., Ltd.)
Chapter 7: Mr. Tadashi Kurosawa (Former Executive Director, Teijin Ltd.)
Chapter 8: Dr. Michikage Matsui (Hiroshima University, previously Director, Osaka Research Center, Teijin Ltd.)
Chapter 9: Professor Yasuo Yamane (Hiroshima University)
Professor Katsuhiko Takahashi (Hiroshima University)
Chapter 10: Professor Kunihiro Hamada (Hiroshima University)
Emeritus Professor Kuniji Kose (Hiroshima University)
Mr. Kenji Kawano (Tsuneishi Holdings Corporation, Tsuneishi Shipbuilding Co.)
Chapter 11: Dr. Yukio Hosaka (Satake Corporation)

We must also acknowledge that the graphs and bar charts and some other figures within this edition have been redrawn (with permission) using the commercial software OriginPro 8.6.

The original Japanese text contains formal references at the end of each chapter to source materials in Japanese. A decision was made to omit these references from this English language edition because English language readers would not understand them. That is why there are no end-of-chapter reference lists that might be expected in a book such as this. However, within the text some of the original Japanese works are cited (in translation) by way of acknowledgment. For the same reason, this book does not include a bibliography. The chapters of the original Japanese book were written with a background of Japanese literature. To invent an English bibliography was felt to be false. We hope readers can accept this work as a stand-alone account of a Japanese view.

Yasuo Yamane
Graduate School of Engineering
Hiroshima University, Japan

Thomas H. C. Childs
School of Mechanical Engineering
University of Leeds, UK

1

Manufacturing Industry

This book is about the broad range of people, organizational issues, and cultural issues that affect the character and success of manufacturing companies, particularly when the companies operate across national boundaries. It is written from the point of view of Japanese companies' experiences both at home and in setting up activities overseas, particularly in South East Asia.

The people with whom it is most concerned are the workforce. This first chapter describes some of the technical activities carried out by the workforce and the skills needed by them. In Sections 1.1 and 1.2 the activities and skills needed in a machine tool manufacturing company are described, as an example. How manufacturing practice has evolved with time is the subject of Sections 1.3 and 1.4. It has led to changed activities and skills. The classification of companies by the type of product they make and why different types require different activities and skills is the topic of Sections 1.5 and 1.6. Section 1.7 emphasizes the evolution of abilities and skills from know-how to know-why and the importance of both. How to quantify skills, teach and measure them, and the importance of these both for the development of the workforce and for a company's ability to operate effectively are the subjects of Chapters 2–4.

1.1 THE MACHINE TOOL MANUFACTURING PROCESS

What is involved in manufacturing machine tools, for example turning, milling, or drilling machines, is taken as an example here. The reasons for choosing machine tools are that they are so-called mother machines, indispensable in manufacturing industry. Manufacturing them requires a high level of precision. Many skillful engineers and technicians are

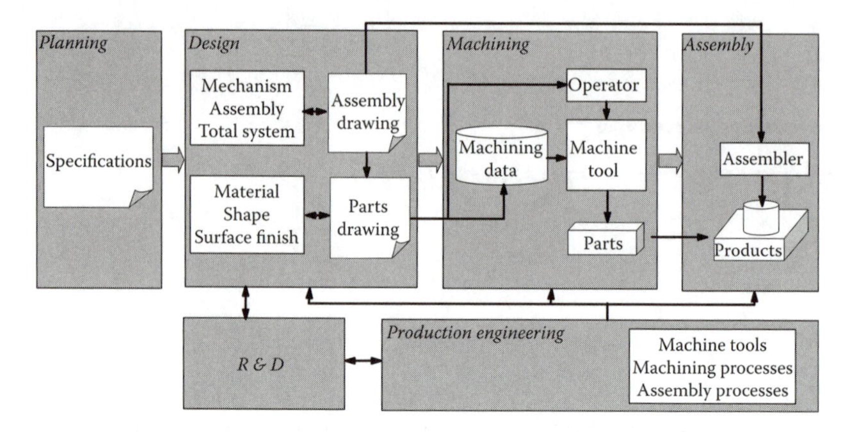

FIGURE 1.1
An example of a process flow in general manufacturing industry.

needed. The manufacturing process for a machine tool is explained briefly before considering the information and object flows that accompany it.

Figure 1.1 shows an example of a process flow, from specifications to products, in a general manufacturing industry. The manufacturing process can be divided into six stages based on function. They are planning, research and development (R&D), design, production engineering, machining, and assembly. Design, production engineering, machining, and assembly will be discussed in this chapter because an evaluation of technical skill is relatively easy in these processes.

1.1.1 Design

In design work, an assembly drawing, which gives an overview of the product, is created first. Next, a drawing of each and every part is produced, based on the assembly drawing. Once the assembly drawing and part drawings are approved, they are turned into work orders and handed over to the divisions that will use them. In days past, designers drew their ideas of three-dimensional (3D) products and parts on two-dimensional (2D) paper using writing materials and according to particular conventions and standards. It was a special skill of designers to copy ideas in 3D onto 2D paper and to reconstruct the information on 2D paper into 3D ideas. It was a skill that was much sought after.

Today, however, designs using computer-aided design (CAD) have become mainstream. It has become possible to create 3D parts virtually in a computer, directly from 3D data, and to combine them to create a virtual 3D product. Movement of and interference between the parts can be checked. If no problem is found, blueprints can be created automatically or data can be produced for processing purposes. Nowadays the old 3D-to-2D conversion skills are hardly needed. Yesterday's drawing boards and paper are hardly seen in design departments, as the changeover to CAD is almost universal. However, at the same time, electrical controls are so frequently used for machine tools that design is often divided into electrical control design and structural design. Now, the ability to integrate these disciplines is much sought after.

1.1.2 Production Engineering

The production engineering part of the manufacturing process is not generally well understood by people outside manufacturing industry. The information on drawings from the design department is normally not sufficiently complete for the drawings to be passed directly to the processing and assembly sites. The information includes images of the completed parts and products, what materials and processing methods (for example milling, drilling) should be used, and dimensional and accuracy requirements. In order to complete the work according to the design drawing, it is necessary to decide in more detail on the procedures relating to manufacturing. For example, it must be decided which machine tools to use, how to hold parts in the tools during manufacture, in what order to make the different features, and when parts should be inspected during the manufacturing process, all taking cost into account. It must also be decided in what order to assemble the parts and, depending on the situation, whether it is more effective to make the part in-house or to order it from an outside supplier.

In the past, the design department would check all these. However, as the complexities and options in manufacturing have increased, now the design department creates the ideal image of the product. The details of manufacture and inspection are left to the experts. The production engineers in the production engineering department are those experts. Making a successful product depends much on this department.

1.1.3 Machining

Machining is a processing department. Machine tools are either so-called manually controlled machine tools or numerically controlled (NC) tools. A manually controlled machine tool, as shown in Figure 1.2, is one that

FIGURE 1.2
A manual lathe. (From Okuma Corporation, model LS. With permission.)

a worker controls manually by turning handles and levers to select cutting speeds and tool movements. Consequently, the quality of each part made in a manually controlled machine tool is directly affected by the worker's skill. On the other hand, cutting speeds and tool movements of an NC machine tool, as shown in Figure 1.3, are controlled by servomotors. The actions of these are controlled by programs created in advance, called NC programs. Once a program is written, the machine tool can be used repeatedly to make many nominally identical parts. The worker's skill becomes setting up and maintaining the machine tool's actions.

1.1.4 Assembly

Assembly is the department where all the machined and other parts are brought together to be assembled. Assembly is divided into subassembly and general assembly. A machine tool (the subject of this example) may be considered to be made up of several components according to their functions. Each component itself has parts. The parts of each component are normally assembled first (subassembled) before general assembly. General assembly is the final process whereby subassembled components are built up. Adjustment and alignment are essential works for giving accuracy and quality assurance (QA) to the machine tool. In the assembly shop, various jigs are used. These are tools for holding parts in position. They play an auxiliary role in helping assembly to be carried out effectively. Furthermore,

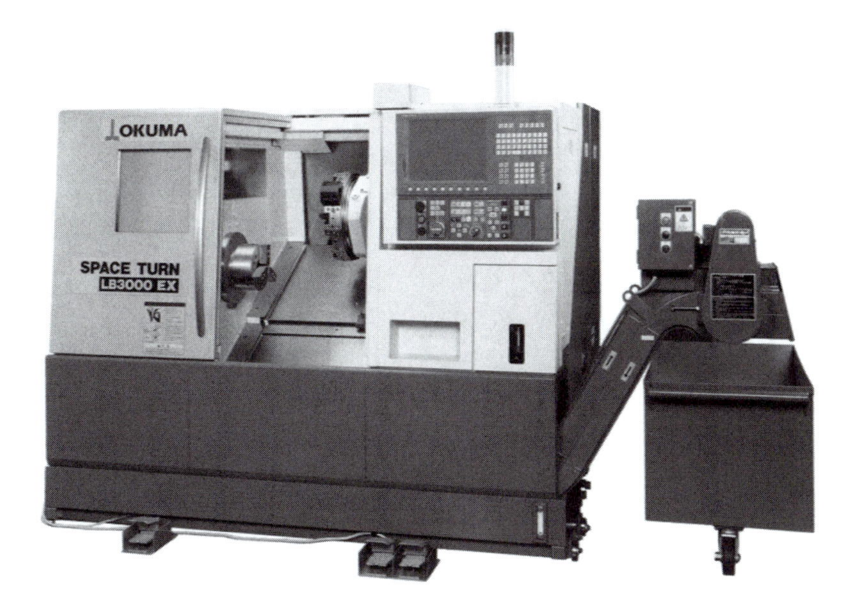

FIGURE 1.3
A numerically controlled (NC) lathe. (From Okuma Corporation, model LB3000EX. With permission.)

a process called scraping is done. This is part of the final adjustment for attaining best possible shape and mechanical accuracy of the assembly. Workers, using a hand tool called a scraper, scrape away small amounts of material from the surfaces of parts that are to be fixed together until the parts fit together with least possible distortion. Normally just a few micrometers in depth are scraped away. Scraping is described further in Chapter 4. It is a most difficult skill to learn.

1.2 INFORMATION AND OBJECT FLOWS IN MANUFACTURING

The stages of manufacturing a machine tool have been explained in outline in the previous section. The accompanying flow of information and materials (raw materials, parts, products) is shown in the block diagram of Figure 1.4. The solid lines show the flow of materials and the dashed lines show the flow of information. To explain the figure simply:

- Planning the purpose of a product, from an idea, is carried out first.

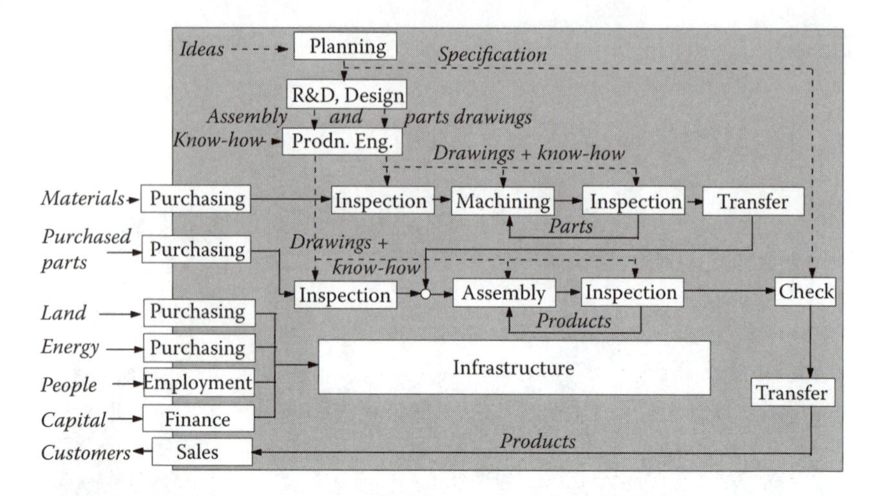

FIGURE 1.4
Information and materials flow in manufacturing.

- Design is then carried out based on the planning.
- The design is developed for manufacture in production engineering.
- Raw materials and parts are purchased and inspected.
- Raw materials are passed on to the machine shop to be processed according to machining information from production engineering.
- The machined parts are inspected and any defective ones are returned to the machine shop.
- Parts that have passed inspection are assembled in the assembly shop.
- Finished products are inspected and if a defective unit is found, readjustments are carried out.
- After inspection, the products are sold and sent to the customer.
- In addition, a department for preparing land and energy facilities as well as sufficient infrastructure is important for these actions to take place.

Although the overall information and materials flows are from upstream to downstream, from planning to assembly, they are not constantly in one direction. Parts are inspected along the way. If a fault is found, a part is returned to the point where the fault occurred and corrections are made. Or the cause of the fault is corrected. Such corrections may be considered to be the result of feedback. Finally, although the flows are shown as occurring sequentially in Figure 1.4, in reality all stages are occurring simultaneously, with different products at different stages of their manufacture.

1.3 COMPATIBLE MANUFACTURING METHODS

The mass-produced industrial products in our daily lives rely on compatible manufacturing methods for their production. Products consist of parts that need to be assembled. Compatible manufacturing methods are defined as ones that produce parts within the required dimensional range or precision for assembly. When out-of-range parts are excluded, it is guaranteed that there is no problem with assembly.

In manufacturing, it is important to understand that it is impossible to make a batch of parts exactly to the same size. This is true whatever the manufacturing process, whether it be a hand or a machine process. Although one cause is variability in the materials from which the parts are made, the greatest reason is that the thermal and mechanical conditions cannot be completely controlled when processing the materials into the required shape. For example, in machining work, tool wear is inevitable and causes the part size to differ from that required.

One way to cope with size variations is to carry forward all parts to the assembly stage and then make any small corrections needed for them to fit together. In that situation, every finished part may be different from every other. Also, it needs skill. There is a case from the 1800s (described in Samuel Lilley, *Men, Machines and History*, New York: International Publishers, 1966) in which 200,000 mass-produced muskets had faults and they could not be used because no one could fix them.

If this type of problem is avoided, by keeping size variations to within a range of approximately the same dimensions that leads to no faults on assembly, with no need for reprocessing or readjustment, then the time for assembly becomes much shorter. In addition, if a part of a product becomes broken, it is repairable by replacing only the broken part with a new part. And if repairs can be made only by exchanging parts, less skill is required. Based on this sort of reasoning, in the early 1800s the idea was born to determine what were the permissible size variations that did not interfere with assembly (and the bigger the permissible range, the easier the manufacturing process). Limit gauges (or go/no-go gauges) were invented to control what were acceptable parts. Figure 1.5 shows an example of a go/no-go limit gauge, used to check whether an inside diameter is within a certain range, in this case between 100 and 100.05 mm.

Production of parts within controlled limits led to a reduction of skills needed and time taken for assembly. Mass production of products that

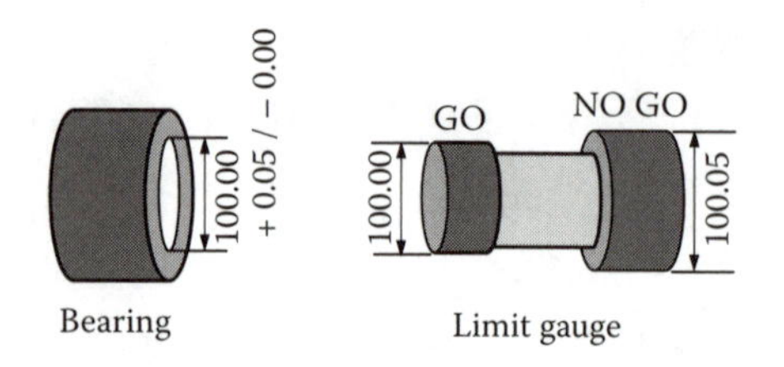

FIGURE 1.5
A go/no-go type of limit gauge.

were made up of many parts (for example machines and clocks) became practical. As a result, prices were greatly reduced. The progress of manufacturing based on compatibility of parts led to the manufacture of the Ford Model T in the 1900s. Thus, the idea of compatible manufacturing methods grew.

1.4 CHANGES IN PROCESSING ACCURACY

In order for the idea of compatible manufacturing methods to be effective, it is important that the number of defective parts, that do not meet the required standard of accuracy, is kept as low as possible. Excluding defective parts by inspection is one possibility for making assembly simple. Decreasing the number of defective parts in the first place, by improving the parts' processing precision, is a better solution. Improving the precision capability of the machine tools that make the parts is essential for this purpose. The precision capability, or accuracy, of machine tools has improved year by year, as shown in Figure 1.6. The figure also shows the precision capability or accuracy of measurement advancing ahead of these. (Precision capability and accuracy have different meanings for a measurement expert, but accuracy is written in the figure, and used hereafter, for brevity, and it is easily understood by general readers.)

The figure's starting time is the Industrial Revolution. Machine tools defined as just for processing (applying heat and force to a material to change it to the desired shape) can be traced back to very early times. However, it was the "copying principle of machine tools" whereby "the accuracy of the processed parts depends on the accuracy of the machine tool" that was established during the Industrial Revolution. Machine tools started to be designed for a required accuracy. This was a clear departure

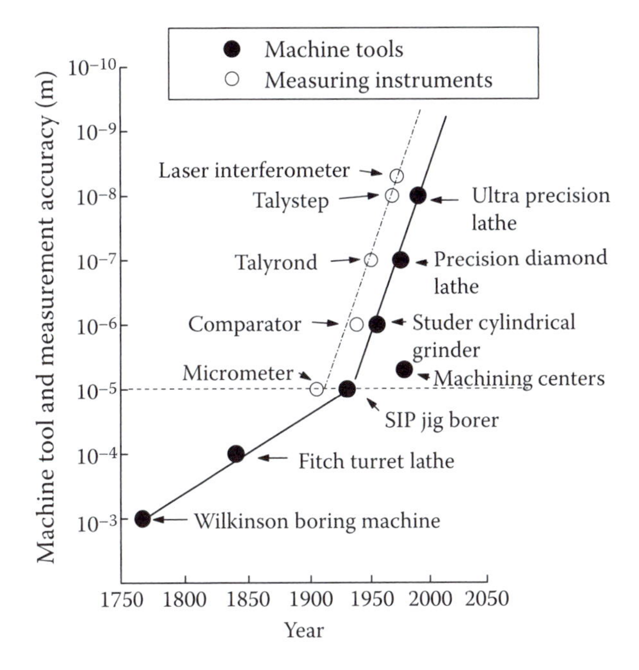

FIGURE 1.6
The historical development of machine tool and measurement instrument accuracy. The machine tool data were originally in M.E. Merchant (*American Machinist*, 103: 142, 1959). They were updated and the measurement instrument data added by Y. Fujimura and T. Yasui (*Machine Tool and Production Systems* (in Japanese), Tokyo: Kyoritsu Shuppan, 1985).

from earlier times. Watt's famous steam engine appeared as a result. Today, as the accuracy of machine tools has improved in every generation, the stage has been reached where accuracy at the nanometer level can be achieved if it is needed.

But 10^{-5} m (0.01 mm) is a critical level of accuracy. The main reason is temperature stability of the machine tool itself, although there may be influences from tool wear, etc., as mentioned previously. Machine tools are mostly made of steel and cast iron. The coefficient of linear thermal expansion of these metals is of the order of 1×10^{-5} per °C. Thus, for machine tools with linear dimensions of about 1 m (1000 mm), a change in size of 0.01 mm occurs when temperature changes by 1°C. This is something that cannot be avoided in general purpose factories. Even though the time needed to process parts may be short, the temperature will change during processing, and it will influence the part's dimensional accuracy. In cases where accuracy below a few micrometers (10^{-6} m) is necessary, the machine tool itself needs to be placed in a temperature-controlled environment. In general purpose factories, there is as little temperature management as possible because of the increased facilities cost. Hence, in cases where processing accuracy of 0.01 mm or below is needed, increased costs are expected and need to be taken into account.

1.5 CLASSIFICATION OF MANUFACTURING INDUSTRY AND PRODUCTS BY NUMBER OF PARTS AND PROCESSING ACCURACY

The larger the number of parts there are in an assembled product, the more difficult the design becomes. From a manufacturing point of view, the higher is the required accuracy of a part, the more advanced must be the manufacturing process. Therefore, the characteristics of a product may be classified by the number of parts in it and the processing accuracy demanded. Figure 1.7 shows an outline of the relation between number of parts and processing accuracy that are associated with a range of industrial products. In the case of processing accuracy, industrial products are made up of parts of various accuracies. Their positions in Figure 1.5 are determined by their main parts. For example, parts with the highest accuracy are required for personal computers (device industry). The processing

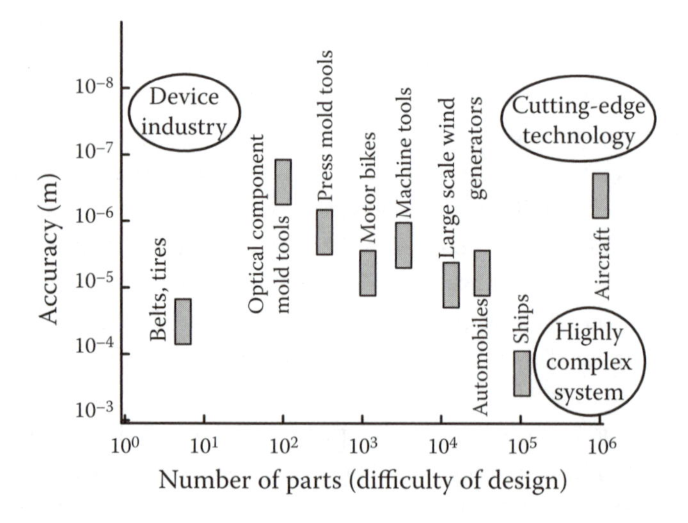

FIGURE 1.7
Industrial product classification by number of parts and processing accuracy.

accuracy of integrated circuits (ICs) reaches the order of nanometers. But this does not mean that the accuracy of the product is in nanometers.

The type of market of a product may be characterized by its position in the figure. For example, in the bottom left of the figure are rubber products such as tires and belts, made of only a few parts, with low accuracy requirements. There is not much difficulty in their design and manufacture. Hence, not much added value comes from making them.

In the top left of the figure, there are products with a small number of parts but which require very high processing accuracy. An example is IC material wafers made from super-high purity silicon (99.999999999%) and the devices made from them. Mass production is necessary to spread the cost of the very difficult manufacturing process. Price becomes controlled by supply and demand.

In the bottom right of the figure are products like cars and ships for which extremely high accuracy is not required but the product has a large number of parts. The large number of parts, involving many suppliers, makes design and development difficult, but the products are not difficult to manufacture. The added value becomes very high.

In the top right of the figure are products such as space satellites and airplanes that require highly complex design and manufacture. (Although a high accuracy is not required of the plane itself, it is required of the engine, navigation equipment, etc.) This type of product has a very high added value but a limited market for each.

1.6 INDUSTRIAL FIELD AND THE TYPE OF TECHNICAL SKILL REQUIRED

Here, the different types of technical skills needed in each of the three areas of highly complex systems, advanced materials processing (device industry), and cutting-edge technology are considered.

Highly complex system industries, such as the auto industry, have two features: an extremely high number of parts and high quantity of production. It is necessary for these industries to employ a wide range of and many skillful engineers and technicians. There is no guarantee that enough engineers and technicians can actually be found to fill every function. Hence, general skills are needed as well as automation of production. Furthermore, because of the huge production, it is extremely serious for a company if products have to be recalled because of faulty parts, particularly if the faulty parts are important ones that affect the whole system. Thus, reliability of both parts and the whole system is most important. Reliability of parts is often simply a matter of their mechanical manufacture, but system reliability involves not only reliability of hardware like parts, but also the reliability of software that allows a part to function as an element of the whole. Although both the architecture and detail of the design are important for the software, it is the basic or architectural design that is most important. It is equivalent to the assembly drawings in machine design. Engineers who can cross boundaries between fields and adjust the technology of one field to another and who can oversee the whole system become indispensable in the highly compex system type of industry.

In the case of advanced materials processing industries such as the device industry, unlike the highly complex system type, not so many engineers and technicians are required since there are not so many parts in a product. However, since these industries supply materials to other industries, the product quality is extremely critical. Specialist knowledge of the materials or products is required. The reliability of the advanced manufacturing equipment, which creates the materials and products, is also most important. Skillful engineers who well understand these equipments are vital to the industry. In this case it is the skills of professional groups of engineers and technicians that are important.

In the area of cutting-edge technologies, there is not much automation, as the production quantity is limited. The unit price of the individually made products is very high. Engineers and technicians with a high level of experience and training are needed, and they may be in short supply.

1.7 ABILITIES REQUIRED BY ENGINEERS AND TECHNICIANS

Continual improvement (kaizen in Japanese) is an essential activity for increasing production efficiency in manufacturing. Therefore, looking objectively at the process is an essential matter for engineers and technicians. Unexpected problems or accidents may sometimes break out during production. Ideally, it is engineers or technicians who have to take measures to prevent problems or accidents occurring, or they have

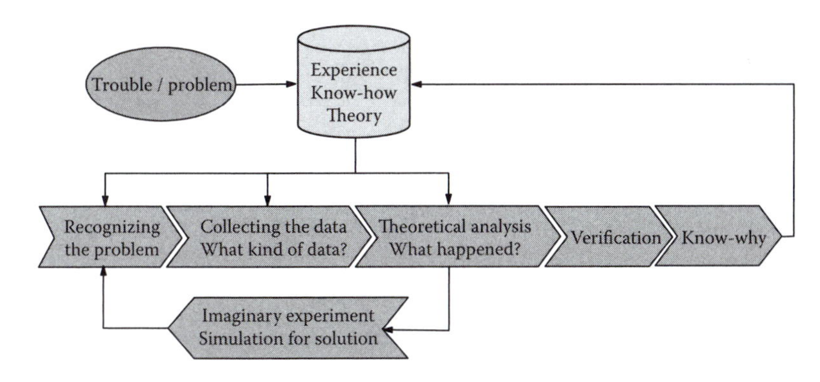

FIGURE 1.8
Abilities required by an engineer.

to take measures to minimize damage from the problems. Problems or accidents can occur frequently, and engineers/technicians have to respond to them quickly.

Figure 1.8 shows a time-series response to a problem. The cylindrical form indicates an engineer's brain. The engineer may deal with the problem as follows:

- At first, the engineer recognizes that a problem is developing.
- Then, the engineer collects quantitative information or data to understand accurately what is the problem. On each occasion, it must be decided what kinds of data should be collected and how to collect them.
- The engineer analyzes the data to speculate what is the cause of the problem, based on theories and experiences.
- "What if" experiments or computer simulations are carried out to confirm the speculated cause.
- Collecting and analyzing data and simulation are repeated until the speculation becomes conviction.
- After coming to a conclusion, verification experiments are devised and carried out.
- If the results of the verification experiments confirm the conclusion, the cause of the trouble can be determined.
- Then the engineer gets new knowledge and experience about the phenomenon that caused the problem. The knowledge can be called know-why because it is based on theories.

FIGURE 1.9
A triumph of hot forging, heat treatment, and grinding know-how: a 76-cm long Japanese sword known as Uesugi-Tachi, Kamakura period, 13th century. (From Tokyo National Museum Image Archives, national treasure.)

Know-why is basically different from know-how. Know-how is knowledge from trial and error. It is one of the important types of knowledge in production engineering, but it is not enough by itself for creating new things. It is knowledge that enables successful repetition of tasks. It is know-why, understanding, that enables invention, and that is the reason why engineers have to engage in it.

Figure 1.9 shows a Japanese sword designated as a national treasure in Japan. The sword was made in the 13th century. Japanese swords were made from Japanese steel called tama-hagane. They were made by a series of processes that roughly may be divided into three parts: forging, heat treatment, and grinding or sharpening. All the processes are very important, but heat treatment, which gives hardness and toughness to the sword, is a quite sensitive and therefore critical process. Heat treatment for cutting tools like swords can be divided into two processes: quenching and tempering. Quenching gives hardness and tempering gives toughness to the tool. Quenching is the process of heating up the tool to a temperature of about 800°C, depending on the carbon content of the tool, and then cooling it down quickly with water or oil. In this cooling process, the cooling rate must be controlled carefully to get a sound microstructure. Tempering is the process of heating up the tool again after quenching. Hardness, toughness, and tensile strength are varied considerably, depending on the tempering temperature, so that temperature is quite important for the tool.

As might be expected, blacksmiths have known since around 1400 B.C., from know-how, that quenching and tempering give hardness and toughness to steel. Understanding the principle of the heat treatment, however, has had to wait until the development of metallurgy during the 19th century, which came from the invention of the microscope and from high-temperature measurement techniques. That is, heat treatment of the steel has been known for over 3000 years, as know-how, but it was only 200 years ago that the operation became know-why.

It is difficult to find a way to solve a trouble or problem only by know-how, especially and almost by definition at the first time of experiencing it. However, if theories relating to the trouble or problem are known, basic principles of a solution can be proposed. Good and successful engineers or technicians must be people with both know-how and know-why.

DISCUSSION QUESTIONS

1. What are the principal functions of design engineering and production engineering?
2. Which judgments would you choose to accept when the judgment of design engineer and that of production engineer differ from each other?
3. It is said that quality of products is quite important. What is quality? What makes the product's quality?
4. Why does a design engineer tend to overspecify a design in working out its details, for example oversizing parts or giving unneeded dimensional tolerances to them?

2

Learning Curves and Their Utilization

How people learn skills and how development of skills may be expressed in terms of a learning curve are the subjects of Sections 2.1 and 2.2. These topics are introduced generally and qualitatively, including by analogy with the behavior of linear systems as described by control engineers. How learning curves may be used both by workers and companies to assess workers' skills and to support workers' development and by companies to evaluate both their workers' and their own technical competence is the topic of Section 2.3. Some industry field studies are reported in Section 2.4. How to define and measure manufacturing skills in particular is the subject of Section 2.5.

2.1 THE LEARNING CURVE

In learning or studying about a subject, it is usual to set an objective and then to start learning or training to achieve it. The clearer and more quantifiable is the objective, the easier it is both to organize a learning strategy and to know how close the objective is to being achieved. Examples of clear and quantifiable objectives are to reach Judo's fifth grade or to reach Calligraphy's fifth grade, or to pass first-level English proficiency test. An example of a less quantifiable objective is to understand Ohm's law. Even in such a case a target level of understanding can be set to test if the learning has been achieved or not.

Generally the learning process has stages, as shown in Figure 2.1. At the start and for a period after that, learning speed is slow. As skills improve, the speed increases more and more. It then slows down again toward the final stage. However, the stages do not have fixed times. If a target is set too

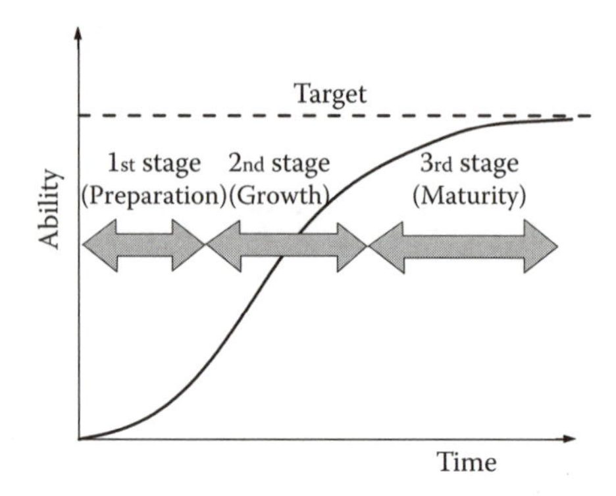

FIGURE 2.1
The learning curve.

high for a particular person, it will never be achieved whatever the amount of effort. If the target is set too low, it may be achieved at the start.

When the target is possible to be achieved, the learning level curve (or learning curve) is expected to show an S shape (as shown in Figure 2.1). The increase of ability as time progress is not uniform. Generally the progress of ability shows the following steps, in order:

- At the start, rapid improvement of skills cannot be expected because time is needed to prepare for learning; for example, ukemi must be learned in Judo before the start of full training.
- After initial preparation, skills become higher as time progresses. However, progress is not necessarily uniform. There can be breakthroughs, ability increasing by leaps and bounds.
- When the target is too low for a person's ability, it is easily achieved.
- On the other hand, when the target is high but not impossible, the progression of skills slows down again after the rapid growth stage, but finally the target is clearly and closely reached.
- However, if the target is set too high to be achieved, it will never be reached.

When considering the handing down or transferring of simple manufacturing skills or manufacturing technology, there may be no problem because such skills or technology are easy to hand down or transfer. However, problems become greater for higher-level skills and technology, when the target level is high enough to be difficult to achieve for most people of average potential ability. Then, long learning and training times and great efforts are needed.

Even if the concept of the S-shaped learning curve (Figure 2.1) is easy to understand, the following questions still remain because learning ability varies from person to person.

- Is it possible to draw a single sigmoid curve as shown in Figure 2.1, for example, for the case of learning design skills?
- What is the use of making an effort to obtain such a progress curve?
- Since the speed of learning and obtaining new skills depends on an individual's potential, shouldn't there be a different curve for each person?

To start to answer these questions, the sigmoid learning curve results from the collection and analysis of a large amount of data. The curve represents the average value of learning vs. time. Continuing with the design skills example, Figure 2.2 charts schematically ability (design ability), on a scale from 0 (minimum) to 1 (maximum), against years of experience. The points show attained abilities, after 5 years, of different people engaged in design work. They demonstrate the common observation of a very wide range of ability achieved by different people with the same years of experience. There are people who may achieve an ability of 0.5 in a short time, while others may take much longer. The S curve represents the average value among all the people. In other words, it is considered that the ability of actual people is distributed around the average value. This qualitative statement may be made quantitative by the use of statistical analysis. The

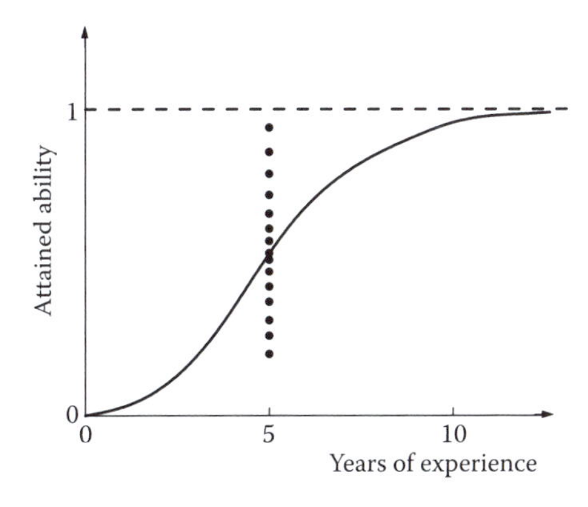

FIGURE 2.2
Dispersion of attained ability.

use of statistics permits an intuitive judgment about a person's speed of learning, and suitability for a job, to be made quantitative. Once it is made quantitative it is possible to say, for example, that a person is not suited to a task. This makes it possible to transfer the person to another job to the benefit of both the person and the enterprise.

2.2 ENGINEERING EQUIVALENTS TO THE LEARNING CURVE

The sigmoid curve describes many phenomena. As an example, the response to a step change of an automatically controlled system having high-order elements is introduced in this section. The response of such a system shows three stages similar to the learning curve's three stages of preparation, growth, and maturity, and these stages can be described mathematically. Furthermore, the input step change signal to the system is analogous to the learning target, since the input signal is the targeted value of the system. However, the purpose of this section is not to explain the automatic control system, but to quantify the several stages in the learning curve, in theory.

Figure 2.3 shows a boiler system for hot water controlled by a person. When, for example, the target temperature is 42°C and the initial, room, temperature is 0°C, the person will open the fuel gas valve to the full, because the difference between the target temperature of 42°C and the temperature from the thermometer is large, and the person will want to

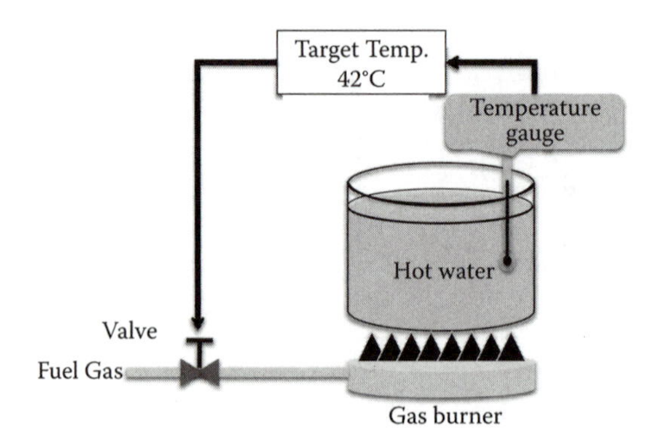

FIGURE 2.3
A boiler system.

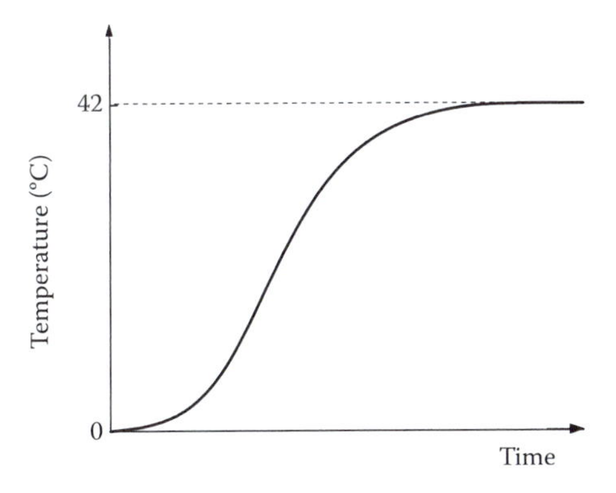

FIGURE 2.4
An example of temperature change.

get hot water as soon as possible. As time advances, the temperature of the water will rise, but in general the rate of rise will not be constant. At the start the water container is at the same temperature as the water inside it. Because convective heat transfer from the container to the water depends on the temperature difference, the heat from the gas burner will flow into the container but not immediately into the water. As the container warms up, its temperature difference from the water will increase and the water will increasingly warm up too, lagging behind the container. Later heat from the container will increasingly be lost to the cooler air outside it, the flow of heat into the water will reduce, and the water temperature will catch up to that of the container and reach a steady value. But that steady value might not be 42°C. Figure 2.4 shows an example of a successfully controlled temperature rise curve in which the fuel gas valve was turned on at the start by just the right amount.

Figure 2.5 shows a flowchart of the boiler system shown in Figure 2.3. The temperature of the hot water is referenced through the thermometer, and the brain judges the valve opening by the difference between

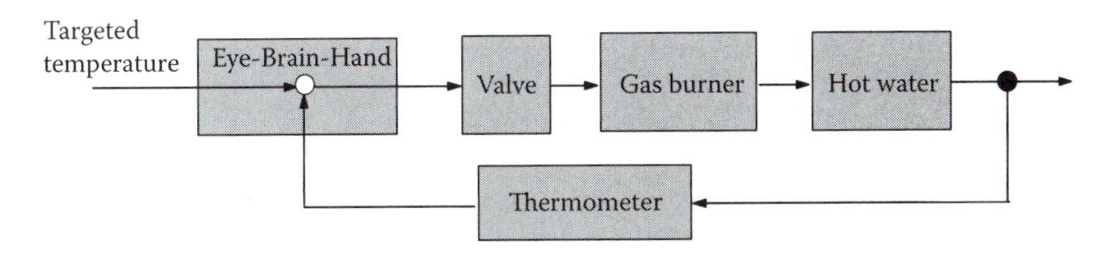

FIGURE 2.5
Flowchart of the boiler system.

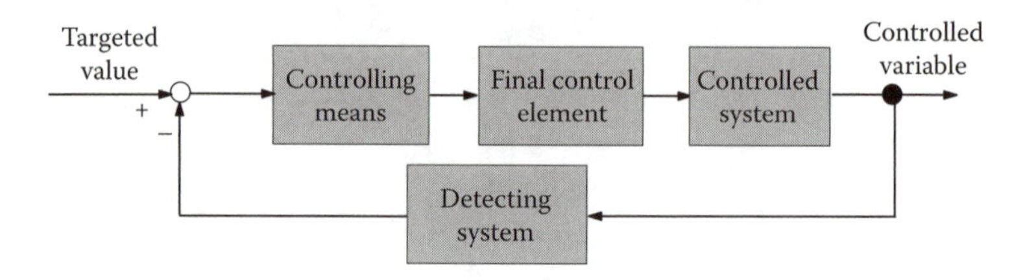

FIGURE 2.6
Basic block diagram of automatic control system.

the targeted temperature and the temperature from the thermometer. Figure 2.6 shows a generalized block diagram of an automatic control system based on this.

- The controlling means corresponds to the person's brain and judges the operation by the difference between the targeted value and the output from the detecting means.
- The final control element corresponds to the person's hands and generates an operating value, such as the rotation angle of the valve.
- The controlled system corresponds to the system to be controlled, such as the hot water tank system.
- The detecting means corresponds to a sensor, such as the thermometer, and detects the controlled variable, such as the temperature of hot water.

Figure 2.7 shows an approximate representation of the response (output) of this system to a step change (input) when the system has higher-order elements. In addition to the boiler system example, other examples are electric circuits consisting of inductance, capacitance, and resistance (LCR circuits) and mechanical oscillating systems consisting of mass, spring, and damper. As shown in the figure, an equivalent dead time, L, and an equivalent time constant, T, can be defined quantitatively by using the tangential line at the point of inflection of the step response curve.

Figure 2.8 redraws the learning curve of Figure 2.1 in the manner of Figure 2.7, by separating the target and ability curves. The similarity between the learning curve and the step response curve is clear, as follows:

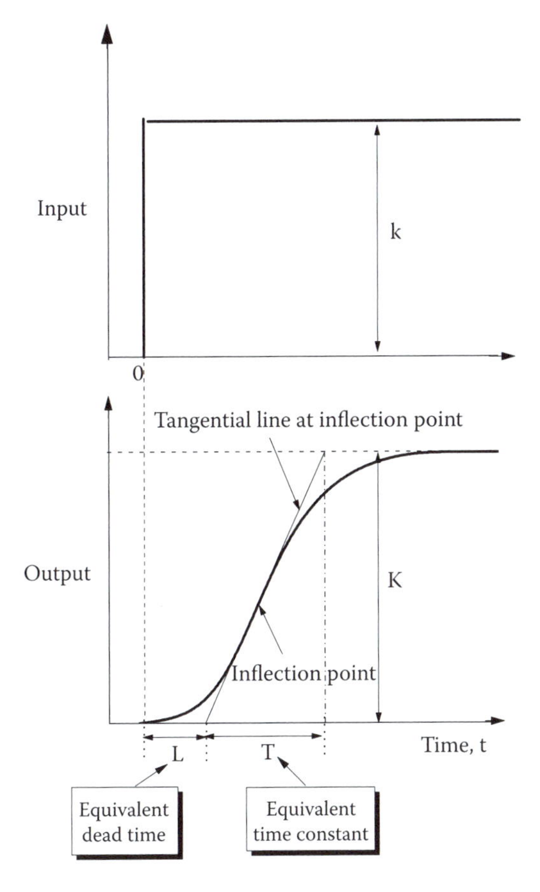

FIGURE 2.7
Approximate representation of step response of a system having higher-order elements.

- The step input corresponds to the target value in learning.
- The step response curve corresponds to the learning curve.
- The equivalent dead time corresponds to the time for preparation in learning.
- The equivalent time constant corresponds to the time for growth.
- The time after the equivalent time constant corresponds to the time for maturity.

By such means the learning curve can be quantified, using the idea of step response curve of a system having a high-order element in the automatic control system.

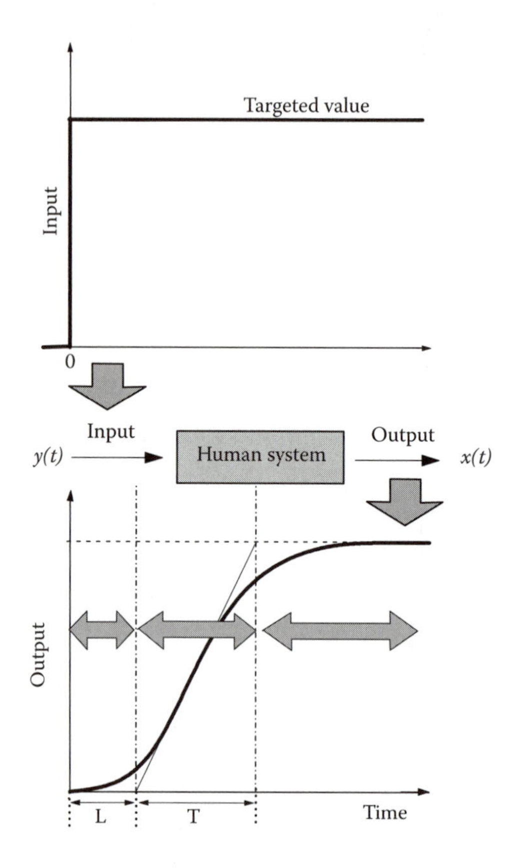

FIGURE 2.8
Functional analogy between the step response and the learning curve.

2.3 SPECIFICATION OF SKILL LEVELS BY MEANS OF LEARNING CURVES

The previous section is concerned with average times to attain skills at different levels. This section is concerned with effects on individuals at one extreme and companies at the other.

2.3.1 Specification of Skill Levels and Its Benefits

Following from Section 2.1 it is supposed that the development of technology and skill in manufacturing takes the form of a sigmoid curve, and that progression along the curve has several stages:

- At first a target to be attained is set (for example to be an expert designer in machine tool manufacturing).
- Then basic training or a preparation period aiming at the acquisition of basic knowledge is needed. How much depends on what is the target.
- Subsequently, on-the-job training starts. When the target is high, learning progress is at first slow. When it is low, it is achieved more easily.
- Once the work is learned to some extent, skill develops more rapidly and a general ability is approached.
- As the goal becomes closer, difficulties increase and the development of skill slows down again.
- In some cases, depending on the situation, development of skill may stop before reaching the goal.
- The higher the goal, the longer the time needed to reach it.

In a company, there are several questions to be answered related to the development of technology and skill, such as:

- How long should be taken for basic training for each different type of work?
- When should one change from basic to on-the-job training?
- How long will be the period of rapid skills increase?
- How long will it take to approach the targeted value after the rapid increase period?

If the development of skills can be quantified, technicians and engineers would be able to assess themselves and know where they stand in their development. This would enthuse them in their work. Besides that, it would be useful to programs of the company's personnel department.

2.3.2 Individual Learning Curves and Learning Curves According to Work

Learning curves can be used either as curves showing individuals' learning speeds or as curves showing learning difficulties according to the type of work. These represent two different viewpoints of human skills. It is natural that skill development differs depending on an individual's abilities, even for the same work. And work itself has different difficulties. Some work can be carried out with a short preparation or training period (for example controlling a simple machine). Other work requires much longer for it to be mastered (for example operation of a skill-oriented machine or production engineering work).

The times to prepare for and master a skill may be considered quantitatively. They can be thought of as the equivalent delay time and equivalent time constant of automatic control system theory, i.e., as the L and T in Figure 2.7 or 2.8.

Figure 2.9 shows examples of learning curves of simple work, skilled work, and in-between work (also called standard work). Simple work, standard work, and skilled work have in general different goals. It is obvious that skilled work sets the highest target. Figure 2.9, however, is standardized, so each work's target is shown as 1.0. But whatever the work, the initial skill level is zero. Figure 2.9 shows all three types of work advancing from a level of 0 toward 1.0, but with preparation time becoming shorter as the work gets simpler.

2.3.3 Evaluation of Individual Skill Levels

When learning curves including dispersion have been created for each type of work, as in Figure 2.2, it can then be judged whether any one worker is developing skills more or less quickly than average. If a worker develops skills much more quickly than the average, it might be for one of two reasons:

- The worker's abilities are perfectly suited to the job at hand.
- The worker has abilities far higher than those required for the work.

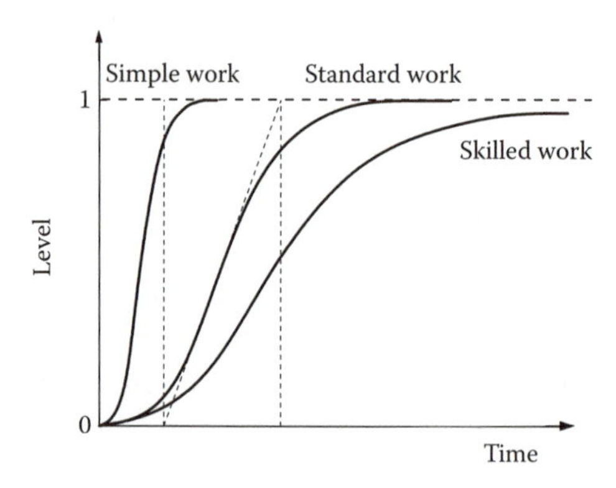

FIGURE 2.9
Learning curves of simple, standard, and skilled work.

In the first case, the worker should be supported and encouraged to continue to advance quickly up the job's learning curve. In the second case, it is beneficial to both the worker and the company to move the worker to a job with a higher level of difficulty.

On the other hand, if a worker shows far slower skill development than standard, the reasons might be:

- The worker's abilities might not match those needed by the job, though the worker might be strong in other areas.
- The worker may just be less able than required for the job.

In both these cases, the worker should be reassigned to a job suited to his or her ability, for the benefit of both the worker and the company. The question also then arises as to when such a decision about reassignment should be considered. If it is delayed too long, disadvantages build up for both the worker and the company. A sensible choice, balancing uncertainty against indecision, is to choose the inflection point time of the learning curve. This is the change point from increasing to decreasing rate of learning and is seen as the point where a beginner moves into the intermediate (or core) skill level range.

2.3.4 Evaluating a Company's Technical Competence

A company's technical ability depends on the sum of the abilities of all the people in it. The total skill level of all personnel becomes the technical competence level of the company. However, the summation of individuals' abilities is not simple. How each individual fits into the company structure must be taken into account. In the case of manufacturing, consider how individual activities are linked together through the block diagram structure of Figure 2.10.

As an example of series connection, consider the partial structure made up of the four divisions planning, design, machining, and assembly. If

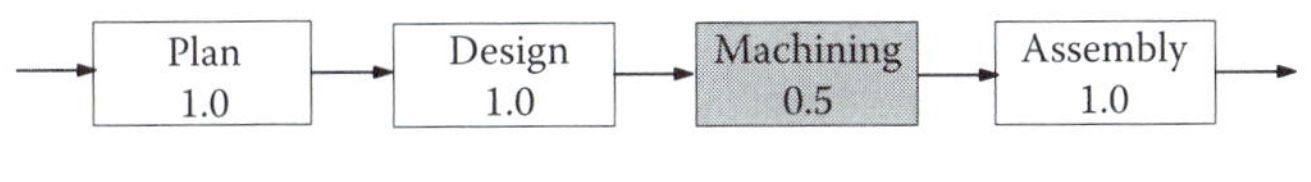

FIGURE 2.10
Total level of series-connected divisions.

each of these four divisions worked at master (1.0, ideal) level, the product made would also be at master level (1.0). However, if, for example, planning, design, and assembly were rated as 1.0, but there were problems in machining leading to faulty parts, so that the machining department's rating was 0.5, then the level of the product would become 0.5. In terms of an equation, if F_P, F_D, F_M, and F_A are respectively the planning, design, machining, and assembly skill levels and F_T is the overall level, then

$$F_T = F_P F_D F_M F_A \tag{2.1}$$

In the case of series connection, which is usually the case in manufacturing, the overall ability is controlled by the weakest ability. If the technical skill level for all divisions does not reach 1, the overall level becomes far from the ideal value. This means that two points are important to raise the overall technical level:

- To increase each individual's skill level as fast as possible.
- To increase the level of the division showing the lowest skill level.

These points will be developed in Chapter 3.

2.3.5 Learning Curves and the Lifetime Employment System

As far as individuals are concerned, their learning curves give information about both their skill level and how long they took to reach that level. Both these factors are the usual inputs that companies and other organizations use to determine workers' pay or other benefits. Three evaluation/reward systems are common:

Type 1: Evaluation/reward based only on individual ability.
Type 2: Evaluation/reward based on service years in an organization.
Type 3: Evaluation/reward based on both ability and service years.

Type 1 applies commonly to professional sports players. Only their current abilities are important to the achievement of their teams.

Type 2 can occur where any of the following apply:

- Effective working depends on the reliability of individuals.

- Understanding of past circumstances is important.
- Familiarity with and an overall view of the organization are important.
- Worker turnover carries risks of leaking of confidential knowledge.

An obvious example is government department employment. Outside government, such evaluation/reward systems, associated with pay dependent on seniority, retirement benefits, etc., are becoming less common.

Finally there is type 3, combining the importance of reliability, understanding company culture, and loyalty (type 2) with current ability (type 1). A strong example is the Japanese lifetime employment system. Recently a point of view has emerged that loyalty (keeping technology and skills within a company) is a most important part of it. But whatever the detail, both a company and its employees would benefit from a quantitative evaluation method. The area under the learning curve provides that. Figure 2.11 emphasizes the area under the curve S:

$$S(t) = \int_0^t F(t)\,dt \tag{2.2}$$

where F is the skill level at time t. S combines both skill level and employment time.

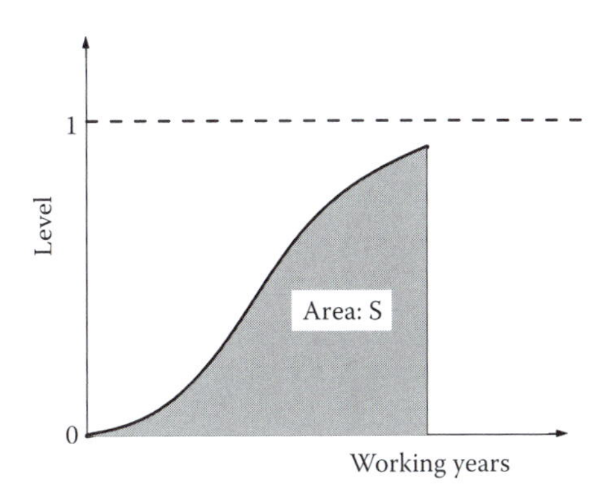

FIGURE 2.11
Individual evaluation using area under the learning curve.

2.4 INDUSTRY FIELD SURVEYS

A number of surveys have been carried out in Japanese industry to estimate times taken for workers to learn skills and the profiles of companies in terms of their workers' skills. Some results are reported in this section.

2.4.1 Skill Levels and Learning Times

A survey to determine skill level scales and the times workers take (on average) to achieve particular levels has been undertaken with five Japanese machine tool makers by one of this book's authors. The survey covered work in all the stages of Figure 2.12: design, production, machining, inspection, and assembly. For a range of jobs in each of these areas it was asked how long it normally took for basic training (i.e., apprenticeship or preparation before starting the job) and how long it took for the following proficiency levels to be reached:

- Beginner: Able to work under supervision/instruction.
- Mid-ranking: Able to work unsupervised except when problems occur; then instruction is needed.
- Expert: Able to work and deal with problems under own judgment.
- Past master: Leads work, able to anticipate and avoid problems.

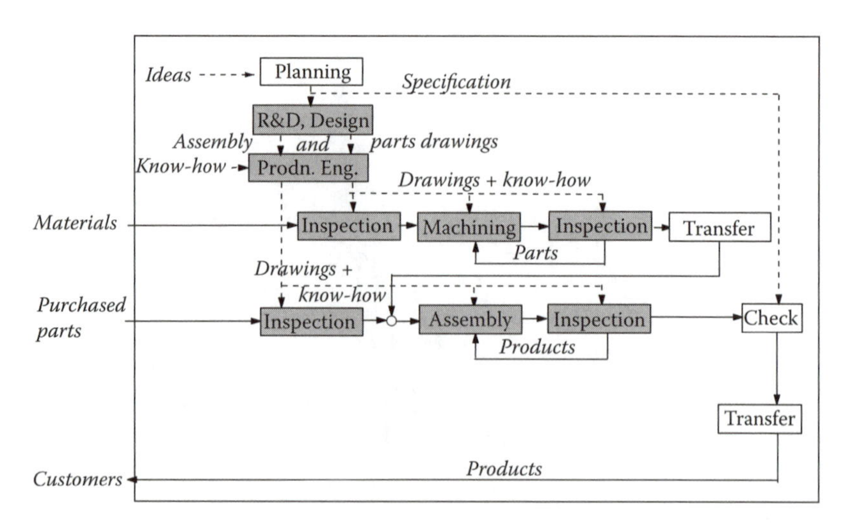

FIGURE 2.12
Information and materials flow in manufacturing.

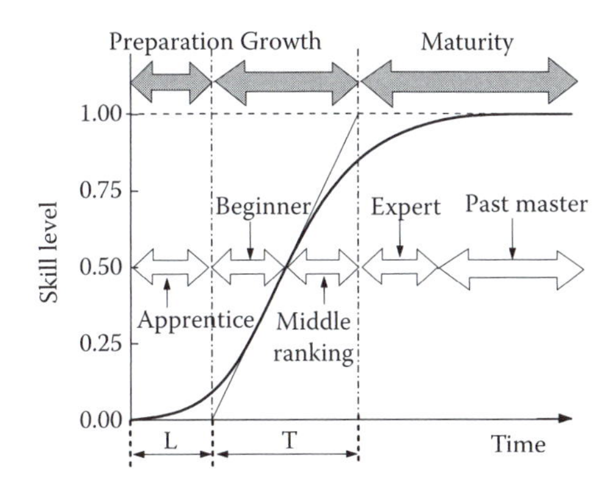

FIGURE 2.13
Skill levels and learning times.

Interpretation of what exactly these descriptions mean was left in the hands of the people in charge of the various sections.

These qualitative descriptions may be made quantitative by converting them to the learning curve skill levels from 0 to 1 with the help of Figure 2.13. Figure 2.13 shows:

- Apprentice: From zero to the start of time constant T (the dead time L).
- Beginner: After preparation to inflection point of learning curve.
- Middle ranking: From inflection point of learning curve to the end of time constant T.
- Expert: From the end of time constant T toward past master.
- Past master: After expert.

With these relationships, from Figure 2.13, the mid-value skill levels for apprentice, beginner, middle ranking, expert, and past master are approximately 0.05, 0.25, 0.75, 0.9, and 1.0, respectively.

Table 2.1 shows survey results from one of the companies (company A). It is the company with the foremost reputation for manufacturing high-quality, precise machine tools.

Figure 2.14 shows the learning curves reconstructed from the answers from the production engineering, machining, and machine design sections. In this case, production engineering and machining have similar learning curves, while machine design is learned more quickly.

The following general points may be made from Figure 2.14 and the original results in Table 2.1:

TABLE 2.1

Years Taken to Attain Skill Levels: Examples from Different Divisions of Company A, a World-Class Machine Tool Manufacturer

	Time (years) to Attain Proficiency Levels				
Division	**Apprentice**	**Beginner**	**Middle Ranking**	**Expert**	**Past Master**
Manual lathe	1	4	14	19	24
Drilling machine	1	4	14	19	24
Manual milling machine	1	4	14	19	24
Manual cylindrical grinder	1	6	16	23	30
Manual cylindrical grinder (internal)	1	6	16	23	30
NC lathe	1	4	14	21	28
NC cylindrical grinder	1	4	14	21	28
Machining center	1	6	16	23	30
Five-axis machining center	1	6	16	23	30
Jig borer	1	6	16	23	30
Painting	3	6	9	12	14
Manual measuring machine	1	4	9	14	19
Automatic measuring machine	1	3	6	9	12
Coordinate measuring machine	1	3	8	11	14
Spindle assembly	1	4	9	14	17
Table assembly	1	4	9	12	15
Feed unit assembly	0.5	1.5	3.5	4.5	5.5
Jigging	1	3	8	11	13
Scraping	2	5	15	20	25
Control box assembly	1	3	8	11	13
Final assembly and adjustment	1	6	16	26	31
Control design	1	3	5	7	10
Machine design	1	2	5	10	15
CAD	2W	1M	2M	0.5	1
Cost computation	0.5	1	3	5	10
Production engineering	3	8	18	23	28

- Among all the jobs of Table 2.1, there is not much difference at the preparation/apprentice stage. The equivalent dead time L is 1 to 2 years in almost all cases.
- Work-dependent differences appear during the beginner and middle-ranking periods. Equivalent time constants typically range from 5 or 6 years up to around 20 years.

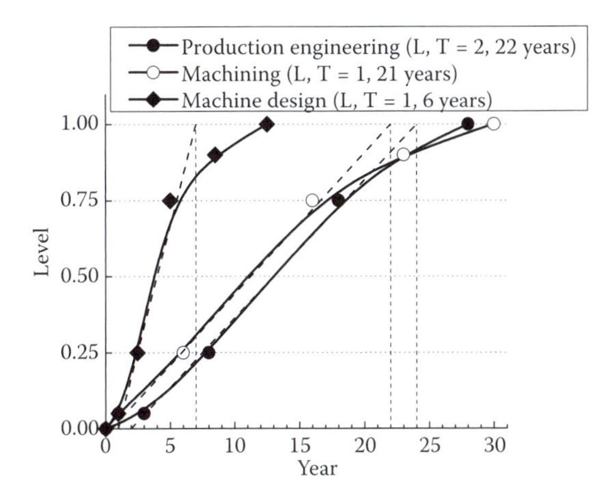

FIGURE 2.14
Examples of learning curves for production engineering, machining, and mechanical design, for company A. The machining data are for machining centers.

- To reach past master level can take from 10 to 15 years (machine design) up to around 30 years (production engineering and machining).

In Table 2.1, final assembly has a similar learning curve to those for machining and production engineering. It takes a long time to learn the necessary fine adjustment skills. As far as machining is concerned, in Section 1.1 it was described that different types of skill are needed for general purpose (manual) and numerically controlled (NC) machining. From Table 2.1 both take similar times to learn. It may be that machining workers have already learned some NC skills in universities or technical colleges or polytechnical institute before entering the company. A related point is that all workers recruited by Japanese machine tool makers will have had previous basic learning experience at a university or technical college. If such times were added to the data in Table 2.1 and Figure 2.14 (and similarly to Table 2.2 and Figure 2.15, next) they would lead to longer L times and more clearly S-shaped learning curves than presented here.

Table 2.2 compares answers from companies A to D for a sample of the divisions of Table 2.1. (Company E did not return complete data. It only gave its own estimates of L + T.) Table 2.2 is similar to Table 2.1 in that each company's learning times for manual and NC machine tool use are the same. Also, the learning times for machine design are similar for all four companies. Otherwise, there are differences. Companies B to D all considered machine design to take a similar or longer time to learn than

TABLE 2.2

Years Taken to Attain Skill Levels: Sample Comparative Data for Companies A to D

		Time (years) to Attain Proficiency Levels									
		Apprentice		Beginner		Middle Ranking		Expert		Past Master	
Division	**Company**	**Data**	**Average**	**Data**	**Average**	**Data**	**Average**	**Data**	**Average**	**Data**	**Average**
Manual lathe	A	1	0.6	4	2.0	14	6.2	19	9.8	24	15
	B	1		3		7		10		12	
	C	0.25		1		3		5		10	
	D	0.02		0.1		1		5		—	
NC lathe	A	1	0.7	4	2.0	14	6.2	21	9.7	28	17
	B	1		3		7		10		12	
	C	0.25		1		3		5		10	
	D	0.05		0.1		1		3		—	
Machining center	A	1	0.8	6	3.2	16	7.7	23	12	30	19
	B	1		3		7		10		12	
	C	1		3		5		10		15	
	D	0.1		1		3		5		—	
Scraping	A	2	1.1	5	3.0	15	9.0	20	13	25	17
	B	1		3		10		13		15	
	C	1		3		6		8		10	
	D	0.5		1		5		10		—	

Assembly	A	1	1.2	6	4.0	16	8.2	26	13	31	19
	B	2		5		7		10		15	
	C	1		3		5		7		10	
	D	1		2		5		10		—	
Machine design	A	1	1.1	2	3.0	5	7.5	10	12	15	17
	B	2		5		10		12		15	
	C	0.25		3		10		15		20	
	D	1		2		5		10		—	
Production engineer	A	3	1.7	8	4.2	18	8.0	23	11	28	18
	B	1		3		5		7		10	
	C	2		4		6		10		15	
	D	1		2		3		5		—	
Overall average			1.0		3.1		7.5		11.5		17.4

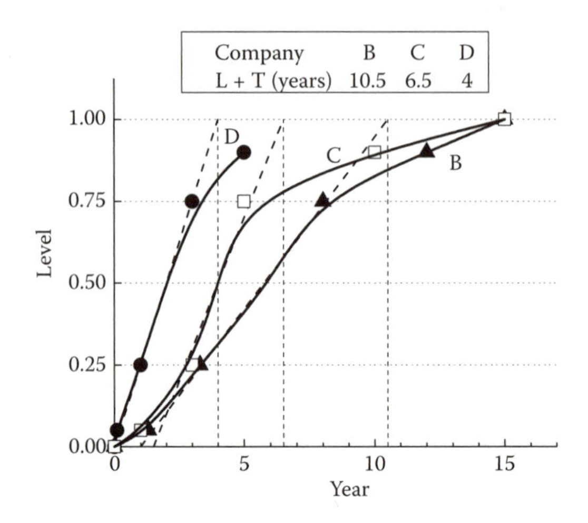

FIGURE 2.15
Learning curves for machining (machining centers), for companies B to D.

production engineering and machining. But more significantly, Table 2.2 shows a wide range of learning times for each of the sample divisions (except machine design).

As an example, Figure 2.15 shows the learning curves for machining center use for companies B to D (the curve for company A has already been presented in Figure 2.14). Additionally, in this case, the L + T time for company E was 8 to 10 years. The range of machining learning times (L + T from 4 years for company D to 22 years for company A, with the other companies from 6.5 to 10.5 years) reflects differences in the companies' positions in the market. Company D specializes in mass production machines where price of the machine tools is an important factor. Expensive, difficult-to-achieve features are avoided. Company A makes specialist machine tools where highest performance and quality are most important. That also explains why company A considers that its manufacturing skills take longer to master than machine design. Manufacturing skills are most important to it. Companies B, C, and E are middle market.

A learning curve may be constructed from the overall average data in the last row of Table 2.2. It is close to the machining center learning curve for company B in Figure 2.15. For the overall data, L + T is 10 years. The times taken to reach the skill levels of 0.25, 0.5, 0.75, and 1.0 are respectively 3, 5, 8, and 17.5 years. This is returned to in Section 2.4.3.

2.4.2 Age and Service Years of Staff

The percentage age distribution of full-time employees in Japanese manufacturing industry is shown in Figure 2.16. The figure's data are based on a survey of 668 companies, published in 2007 under the title *Investigations on Youth, Turnover and Work Commitment*, carried out for the Ministry of Health by the Japan Institute for Labour Policy and Training. The original data were said to be accurate to within 6.7%. From the figure, 72% of the workforce can be seen to be in the age range 35 to 44 years old. The average age is 39.6 years, with a standard deviation of 4.3 years.

The average service years of employees is shown in Figure 2.17. The source of the data is the same as for Figure 2.16. From the figure, the average service years are 13.5 with a standard deviation of 5.1 years. Supposing the service years to follow a normal distribution, from 8 to 19 service years makes up 68% of the whole. It can be concluded from Figures 2.16 and 2.17, although they are based on limited data (from only 668 companies), that the main workforce in Japanese manufacturing industry is in the age range 35 to 44 years old, with service years from 8 to 19 years. (In both figures, the total percent is less than 100. Approximately 5% of the companies did not reply to the survey.)

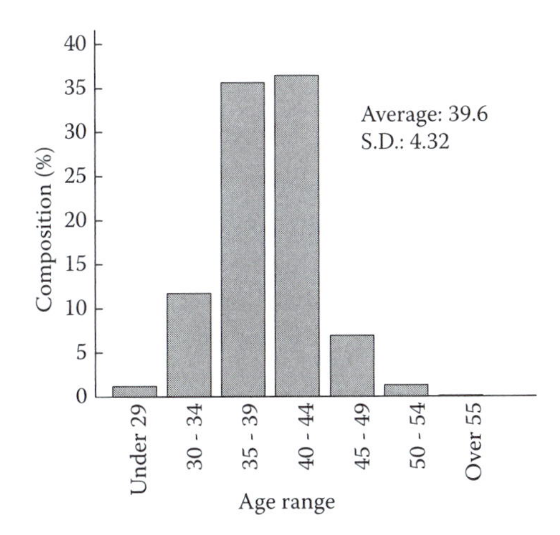

FIGURE 2.16
The percentage age distribution of full-time employees in Japanese manufacturing industry. (From data in *Report on Investigations on Youth Turnover and Work Commitment*, Japan Institute for Labour Policy and Training, 2007.)

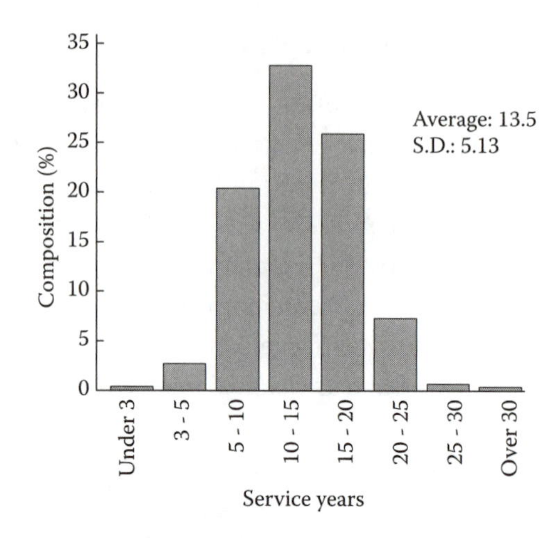

FIGURE 2.17
The average service years of employees. (From data in *Report on Investigations on Youth Turnover and Work Commitment*, Japan Institute for Labour Policy and Training, 2007.)

2.4.3 Companies' Technical/Skill Level Estimated from Service Years

Tables 2.1 and 2.2 present survey data of the times taken on average for workers to reach levels of skill for a range of jobs in machine tool manufacture. The skill levels are converted to numerical values from 0 to 1. Example learning curves from this data are presented in Figures 2.14 and 2.15. This kind of information may be combined with the service years information in Figure 2.17 to obtain measures of the technical level in Japanese company departments or whole companies, as follows.

As listed at the end of Section 2.4.1, the overall average times for companies A to D to reach skill levels of 0.25, 0.5, 0.75, and 1.0 across a range of job divisions are 3, 5, 8, and 17.5 years, respectively. If it is assumed that the service years of Figure 2.17 apply equally to all types of jobs, weighting the skills levels by the percent of employees with the required service years leads, for this example, to a mean technical level for these companies and across the divisions concerned of 0.8. This can be considered to be quite high, being at the middle-ranking/expert borderline (Figure 2.13).

In company A's case, the learning times for the job divisions in Table 2.2 are typically twice the overall average values. Machine design is the exception. Table 2.1 also shows other divisions where times are close to the overall average values. For the long learning time divisions in Table 2.2, company A's mean technical level falls to 0.5, by the same calculation as in the previous paragraph. The level of 0.5 is at the beginner/middle-ranking boundary in Figure 2.13. Company A is the one with the highest quality reputation. Even in Japan, which is considered to have an excellent manufacturing industry, there can be seen to be difficulties in maintaining excellent machining, assembly, and production engineering departments. It makes it important to find ways to reduce learning times. This is returned to in Chapter 4.

2.5 SKILL LEVEL AND STANDARD DEVIATION

Up to this point, skill levels as in Figure 2.9 have been standardized from 0 to 1. How may absolute values be defined and measured in the case of manufacturing? Language learning, for example, is commonly tested by exams. Proficiency in a language can be defined by the grade of exam being taken and measured by the mark gained in the exam. The existence of the grade and mark make it easier to understand what level of skill has been reached. In the case of manufacturing, in Japan, there is a scheme of proficiency measurement exams to grade and measure workers' skill levels.

2.5.1 Proficiency Measurement

Technical training in Japan is governed by law, the human resources development law. Established under that are Japan's Vocational Ability Development Association (JAVADA) and the National Trade Skill Testing and Certification (NTSTC) system. JAVADA promotes and establishes skills and evaluation standards. Its goals are increasing the skills as well as the positions of working people, as well as contributing to the development of industry. The NTSTC system (which JAVADA administers and for which it sets test questions) is one means of attaining these goals.

Some of the skills tested in the NTSTC system are classified as special grade, grade 1, grade 2, and grade 3, while other skills are not classified, as follows:

- Special grade: Skills required by managers/supervisors.
- Grade 1 and nonclassified skills: Skills required by advanced skilled workers.
- Grade 2: Skills required by intermediate skilled workers.
- Grade 3: Skills required by novice workers.

To be eligible to be tested at the levels of the different grades, workers must have had the following years of practical experience:

- Grade 1: 7 years (before 2003 it was 12 years).
- Grade 2: 2 years (before 2003 it was 3 years).
- Grade 3: 6 months (before 2003 it was 1 year).

The eligibility requirement for grade 1 is mainly to include only highly experienced and core workers, even though the qualifying years have been greatly relaxed from 12 to 7 years of practical experience. There is no specific definition of what are novice, intermediate, and advanced skills, though differences in difficulty of work at each level are to be expected. Testing is by both practical work under exam conditions and written papers. There are cases where students from industrial high schools or technical junior colleges have passed grade 3 exams.

2.5.2 Skill Level and Standard Deviation

What defines the differences between novice, intermediate, and advanced skills in manufacturing is considered further here. One aspect relates to the difficulty of the work that can be carried out. Another relates to the assuredness with which it can be carried out. Assuredness includes such things as within the time demanded, with the accuracy demanded, and with the smallest dispersion. In machining, for example, a worker with a high skill level will make parts with a lower dispersion among them and will therefore make a larger proportion of parts within the accuracy demanded than will a less skilled worker.

For those jobs like machining, for which a worker's output will have a normal distribution of sizes, the fractional probabilities that a measured dimension will lie within ±1, 2, 3, or 4 standard deviations σ from the mean are respectively 0.6827, 0.9545, 0.9973, and 0.99994, as shown in Figure 2.18. More broadly, there is a statistical quality control technique called 3σ management, in which the percentage of parts outside tolerance (rejects) should be kept within 0.27% (i.e., 1–0.9973). Today there is an even more stringent quality control technique called 6σ management. In 6σ management, the percentage of rejected parts should be kept within 0.00034%. In order to achieve such extremely small values, complete process control is needed, from upstream to

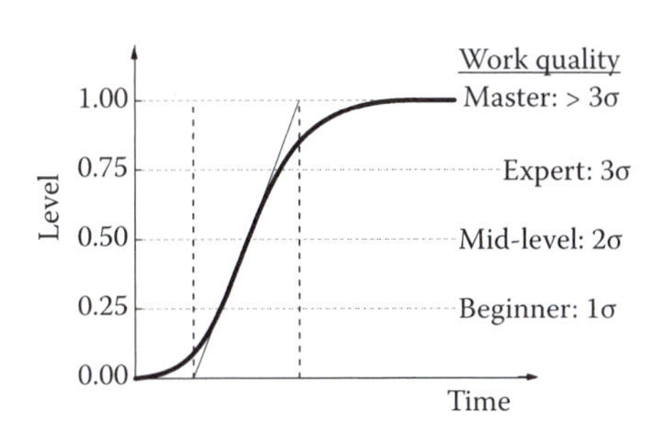

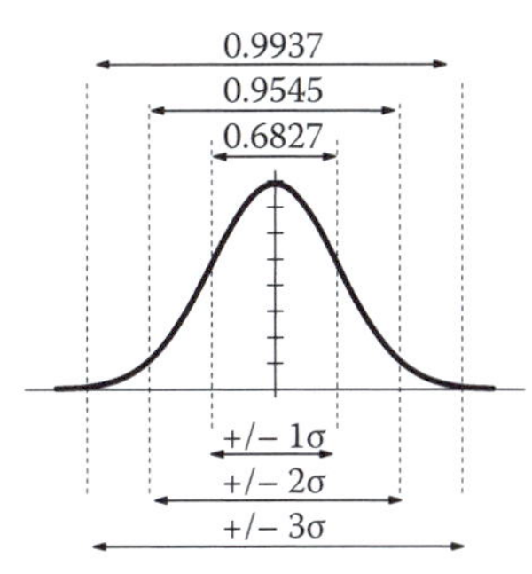

FIGURE 2.18
Skill level and normal distribution.

downstream, from planning to assembly. Especially the inspection process is most important.

Engineers, technicians, and other workers take part in all stages of manufacturing. Their skill levels directly influence the quality of the whole process. Returning to the question at the start of Section 2.5, how to define a skill level scale for manufacturing, it is suggested that for an engineer or technician whose output follows a normal distribution, it could be developed in terms of an ability to manufacture to within a set tolerance to standard deviation ratio, as follows:

- Beginner skill level: Tolerance/σ = 1.0.
- Mid-ranking skill level: Tolerance/σ = 2.0.
- Expert skill level: Tolerance/σ = 3.0.
- Past master skill level: Tolerance/σ in excess of 3.0.

For other types of jobs, for example design or production engineering, similar scales could be constructed, though the definitions of levels would be different. Such scales would exist at all levels, grades 1, 2, 3, etc., of the NTSTC system.

DISCUSSION QUESTIONS

1. Is there implicit knowledge in design engineering? If the answer is yes, give examples, explaining what are the implicit parts.
2. In design engineering, computer-aided design (CAD) technology is developing at high speed. Explain what is CAD technology, and

consider the following proposition: if we can obtain or develop a strong CAD tool, including simulation technology, the design of a product will become almost independent of the designer.

3. A simulator, such as a driving simulator or a flight simulator, is useful to shorten the time needed to learn a skill. What other simulators that support learning can you think of?

3

Skill Transfer in Manufacturing Industries

Chapter 2 makes the general point at its start that it takes time to learn skills. It introduces the concept of a learning curve to help describe a person's progress at mastering a task or skill. Later it is more specifically concerned with the sorts of skills needed in manufacturing industry, defining learning curves of use for tracking workers' progress in manufacturing and indicating the value of learning curves for both workers and companies. In particular, it provides data (Tables 2.1 and 2.2) on the long times needed to learn manufacturing skills. The implication of these long times is that maintaining a highly skilled workforce needs long-term policies. A successful policy must address anticipated needs. One is the shortening of learning times. Chapter 2 also describes the current state of the workforce in manufacturing industry in Japan, to give a background to current Japanese manufacturing culture and the problems that are anticipated. Chapter 3 is about how skills may effectively be handed on and learning times may be shortened. But there are constraints and risks involved. These are outlined too. The overall purpose is to set the scene for later chapters concerned with transferring technology and skills overseas, either under license or to subsidiary companies. It is necessary to appreciate a company's needs and interests in order to understand its actions in transferring technologies overseas.

3.1 TECHNOLOGY AND SKILL TRANSFER

3.1.1 Skill Transfer from the Time of Chuang Tzu

Chuang Tzu was a philosopher who lived from around 360 B.C. to 280 B.C. Figure 3.1 shows his portrait. He wrote the *Book of Chuang Tzu*. One

FIGURE 3.1
The Chinese philosopher Chuang Tzu (4th century B.C.). (From *The Sancai Tuhui*, Chinese, early 17th century.)

way in which he conveyed his thoughts was through stories involving skilled craftsmen, for example carpenters and cooks. In one of the stories, Duke Huan of Qi talks with the wheelwright Bian. The original text is (in translation) as follows:

> Duke Huan, seated above in his hall, was (once) reading a book, and the wheelwright Bian was making a wheel below. Laying aside his hammer and chisel, Bian went up the steps, and said, "I venture to ask your Grace what words are you reading?"
>
> The duke replied, "The words of the sages."
>
> "Are those sages alive?" Bian continued.
>
> "They are dead" was the reply.
>
> "Then," said the other, "what you, my Ruler, are reading are only the dregs and sediments of those old men."
>
> The duke's face changed and he said, "How should you, a wheelwright, have anything to say about the book which I am reading? If you can explain yourself, very well; if you cannot, you shall die!"
>
> The wheelwright said, "I have been making wheels for many years as a wheelwright. If I were to shave the wheel too much, there would be a big clearance between the wheel and the shaft. However, if I were to shave too little, the shaft would not fit into the wheel. Whether I have shaved too much or too little all depends on the senses of my hand. It is impossible to tell (how to do this) by word of mouth. This is gained from experience. I cannot teach this sense to my son and my son cannot inherit what I have experienced. Thus, although I am in my seventies, I am still making wheels in my old age. But these ancient people, what they gained and experienced could not be written and passed down. So, what you, my Ruler, are reading is but their dregs and sediments!"

The thing that Chuang Tzu wanted to convey is thought to be the last part, where the important thing to be conveyed is not something understood in the mind but something understood in the heart (in Buddhism, this is also known as spiritual enlightenment), or perhaps what is mastered cannot be passed down in words. If Chuang Tzu had not experienced the situation as described by the wheelwright, he would not have taken the story as an example.

The shaving problem faced by wheelwright Bian as related by Chuang Tzu was solved about 2000 years later. Although it may be a diversion, consider this problem a little further. Figure 3.2 shows a shaft in a bearing. A little clearance is needed for the shaft to rotate in the bearing, and the clearance must be filled with oil for lubrication. The problem here is what is the optimum value for the clearance. Actually, the answer depends on how the shaft is used. Although there are many points of detail, the following general points can be made:

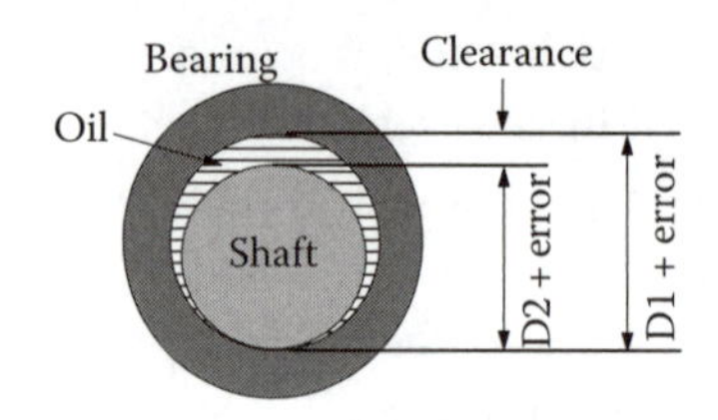

FIGURE 3.2
A shaft and bearing.

- The shaft rotates in the bearing (in some cases the shaft is fixed and the bearing turns) and friction occurs as a result.
- Friction depends on the load carried by the shaft and the rotation speed.
- Heat is generated by the friction, and the temperature of the shaft and bearing increases.
- Generally, the temperature of the shaft increases more than that of the bearing as the heat escapes more easily from the bearing.
- The shaft expands as its temperature rises and the clearance between the shaft and the bearing becomes smaller.
- When the thermal expansion of the shaft becomes extreme, the clearance may fall to zero. This situation is called seizure.
- So the answer to what is the optimum clearance is what is needed to prevent seizure, or maintain the clearance at some appropriate value above zero, under the applied conditions of load and sliding speed.

There are actually two problems to be solved in manufacturing the optimum clearance. The one just considered is what is the relation between conditions of use and clearance. Historically, until the development of bearing lubrication theory, what clearance to manufacture depended on intuition and experience. The other is the problem of inaccuracy in manufacturing. Until the idea of dimensional tolerance was introduced and until it was possible to measure shaft and bearing sizes at the levels of accuracy needed, it was not possible to make use of the knowledge of what the clearance should be.

In summary, while technology was a matter of experience, it was difficult to transfer knowledge. Now that more is understood, it is easier to explain it.

3.1.2 Technical Skill Classification

As considered in the previous section, there are some parts of technical skill that are understood rationally and some parts that come from intuition and experience.

As far as transferring technical skills is concerned, those skills that are understood rationally can be transferred through records such as words; those that come from intuition and experience cannot be. Those that can be conveyed through records are called explicit knowledge, and those that cannot be are called tacit knowledge.

In manufacturing, how to do things and technical standards based on theory, that are learned from records made by people some distance away from the workers, are examples of explicit knowledge. It does not matter whether the records are words or pictures or demonstrations. Teaching in technical high schools and technical colleges or universities is part of gaining explicit knowledge. It goes without saying that the recipient of explicit knowledge must have the ability and background preparation to interpret the record correctly. An appropriate foundation is required to master, for example, theories of lubrication and ideas of dimensional tolerance. Even if a complete record is left behind of how to do something, it is completely meaningless if it is not understood fully. For example, if lubrication theory and dimensional tolerance were to be taught to elementary school students, it would be very difficult. Such students do not have the foundation needed. Equally, it takes effort on the part of the creator of the record to convey it clearly.

The wheelwright Bian example from 2000 years ago is one of tacit knowledge. There are many other examples from historical times, when engineering explicit knowledge was limited and making things depended on a great talent and special understanding between teacher and pupil. One is the manufacture of swords (Figure 1.9). A master could make blades of high quality, able to keep a sharp edge, without today's knowledge of ferrous metallurgy nor instruments to measure and control temperatures in heat treatment. Today there are still examples where tacit knowledge is important. Many are not from manufacturing. Spiritual enlightenment in Buddhism requires a state to be attained from long meditation and guidance from a master. There are sports athletes who have a special talent but cannot explain to others how to perform like them. Closer to manufacturing there are artists, especially potters, who can make beautiful objects without having a technical knowledge, for example, of glazes. They have gained their ability through many years of trials, at first by following a master, later perhaps by creating new techniques themselves. One hundred years ago, manufacturing was called a useful art.

Even today in manufacturing there is a well-known problem of the retirement of a key worker, when only that worker knows how to carry out a special operation. Many processes require special adjustments when they are being set up. From an explicit knowledge point of view there may be many possibilities for adjustment. The key worker may have found a specially effective way to select and carry out the adjustments but cannot explain how he does it. In that case the future ability of the company to carry out the process depends on finding a follow-on worker with special talents who can establish a special relationship with the retiring worker and learn tacitly from that relationship. The human being is a clever and intuitive learning machine and can sometimes create a good result without any formal theory.

In summary, tacit knowledge is knowledge that cannot be passed on in records and by rules. It requires a special talent and understanding between teacher and pupil for its transmission. It perhaps comes as much from intuition as from the rational mind. Its importance in manufacturing is often not appreciated by people who do not have experience of it.

3.1.3 The Teaching of Technical Skills

At what stage of a person's development should what technical skills be taught? Where should the teaching take place? Who should carry out the teaching? These are all questions that are relevant to how to transfer technical skills. The answers will change depending on where a person is on the learning curve: at the induction and novice stage, the fast learning stage or the maturity period. And sometimes the answers may not be certain.

Consider what learning is needed to be able to design a machine tool, a machine at the heart of manufacturing. Some of the things that need to be considered and their related fields of study are:

- The requirements of machining: Machining process theory.
- Materials for the machine tool: Engineering materials.
- Control of the machine tool: Control engineering.
- Machine tool vibration: Dynamics of machinery.
- Deformation under load of the machine tool: Strength of materials.

- Thermal stability of the machine tool: Thermodynamics.
- Lubrication and cutting fluids for the machine tool: Hydrodynamics.
- Usability of the machine tool by workers: Ergonomics.

There are other, more general topics to be considered, such as engineering ethics and law. And all must be brought together in the designing process. There is mechanical design theory too.

These fields of study, engineering materials and so on, are well understood at a basic level. They are examples of explicit knowledge and are taught in higher education, at technical colleges and universities. The start of the learning curve is usually taken to be after this basic learning period, once a person has joined a company. For example, although mechanical design theory may be taught in a technical college or university, the particular design methods and procedures that a company uses for machine tool design will be too specific to be taught as part of general education. Once an employee enters a company, the know-how of machine tool design, and starting on the learning curve, will be taught by senior staff during on-the-job training.

At this early stage of learning it does not matter if the training people are mid-level personnel or veterans. It is just important that they are good teachers. Not all skillful engineers or technicians make good teachers. If they have taken 20 years or more to reach their present positions, there will be a gap of 20 years between their backgrounds and the newcomers' backgrounds. There may have been large changes in the basic fields of study in the meantime.

What should be taught at the early stage may be answered, staying with the example of design. Although there are many things to learn in order to design something completely new, with no previous examples for guidance, designing something completely new is unusual. Normally designs are incremental. They are developed from previous designs. The important thing is for the teacher to help the learner to understand why the earlier machines were designed as they were, based on the information in the old design drawings. The teacher should not give the answer from the start. The skills that he should try to develop in the learner are such as:

- Having a questioning and critical mind.
- Having the ability to apply experience and knowledge to solving problems.
- Searching for answers.

- Being able to verify an answer.
- Becoming confident.

These are not just the answers to questions. They include the process of thinking, which consists of the process of recognizing the problem and looking for the answer. It is important in all aspects of manufacturing, not just design.

After acquiring some initial know-how, a worker enters the rapid growth period. Once he has become sufficiently reliable on the job to be considered independent, he enters the maturity or expert period. Although all may agree that know-how is very important, there is a question whether it should be considered as explicit or implicit knowledge. There is an opinion, in the case of machine tool design, that more than half of know-how can be conveyed through records; i.e., it is explicit knowledge. However, in this example, where the record is mainly drawings and related material, there are problems of both loss of records and maintaining a huge library of records, requiring an archiving office. It may be more effective to pass down know-how implicitly, from person to person, even if a record does exist.

Who should be the teachers of mid-level workers, in order to help the workers become expert? What should they teach, how, and when? Those qualified to teach are those who are already well established as experts. The opportunities to teach come during work when problems arise that the mid-level worker cannot solve for himself.

In this kind of circumstance, rather than passing on know-how in detail, it is generally more effective just to give some hint or advice about considering the basic cause of the problem, or even to use stories in the manner of Chuang Tzu. However, the really good teacher will recognize that occasionally it is better to be more direct. If trouble can be anticipated beforehand, it might be better just to say "stop."

People passing on skills to other people has been a feature of humanity in all walks of life, from time immemorial. There is still no simple answer about what is the best way. There are many kinds of people, both teachers and learners, and there are many things to be taught. All these make it difficult for a simple answer to exist. At least being aware of the difficulties and analyzing the special needs and purposes of manufacturing skills transfer should help make the teaching of manufacturing skills more effective.

3.1.4 Learning Curve Time Reduction

It would be of great benefit to manufacturing companies if the long learning times, for example in Tables 2.1 and 2.2, could be shortened. In the same way as in the last section, problems and possibilities of shortening learning times can be considered for each of the early, fast learning, and maturity stages of the learning curve.

3.1.4.1 The Early Period

How long the early period takes depends very much on a worker's basic knowledge before entering the company. Problems can arise from not being clear what is and what is not included about manufacturing in the engineering departments and technical colleges of the higher education system.

Companies' views of what should be included are different now from in the past. Companies used to consider that technical things can be learned after joining the company; the important thing in higher education is to learn the basics and mathematics. Now, because of economic pressures, although some companies have not changed their view, many now ask for talented people who can step in and be effective immediately.

From the universities' and technical colleges' points of view, basic skills and effectiveness are two different things. The structure of a curriculum for developing basic skills is different from that for achieving effectiveness. Further, higher education does not exist only for industry but to educate students more widely for the benefit of society as a whole.

However, formation of people able to take their place in industry is certainly one of the goals of higher education. The transition between education and early period company training, taking into account the companies' views on how to be most effective in early training, needs the collaboration of both parties.

3.1.4.2 The Fast Learning Period

Because the fast learning period is already associated with a steep learning curve, it is not easy to see how it may be significantly shortened. However, that does not mean that shortening it is impossible. The detailed skills level survey data in Table 2.1 of Chapter 2 has given examples of learning times shorter today than in the past as a result of advancement of technologies.

One example is in the area of design. Spatial reasoning ability used to have to be acquired, in order to represent 3D parts by 2D drawings and to visualize the 3D content of the drawings made by other people. Companies reported that today 3D computer-aided design (CAD) makes it easy to alternate between 3D and 2D views. This has led to a reduction in the design skills learning time. Another example is machine tool operation. In the past, workers had to learn how to grind their own cutting tools and set them in their machines. Today's machine tools (both manual and numerically controlled (NC)) and tooling systems mainly use pre-prepared "throw away" tooling inserts. The learning period has been shortened through changing what needs to be learned. It is believed that new technologies, not necessarily manufacturing technologies but education technologies, should be promoted strongly for their potential to reduce fast learning period times. They can change the way in which skills are learned. More on this is explained in Chapter 4.

There is one other point to be made. Experienced workers have often brought their own understanding to a job. As a result, they have developed their own improved way of doing it. By observing a senior's work and thinking about why or how it may be a better way to do things, or how it may avoid problems, someone moving up the fast learning curve may in fact accelerate his own development. Looking for one's own answers is a skill learned in the early period. It is certainly important in the fast learning period.

3.1.4.3 The Maturity Period

Once a worker has become an expert, becoming a master becomes the goal. Self-motivation is important for a worker to continue to improve. General ways to shorten the learning time in this period become not so important. Rather, in addition to increasing skill in his own work, consideration should be given to two other aspects of how the worker can contribute to a company's performance. These are how to make more effective his skills at supervising his juniors and how to ensure his skills remain of value to the company in the event that he moves to a higher management job.

Passing on the culture and tradition of technology and skills from masters to juniors has been how companies have transferred skills in the past. This is now breaking down (this is known as the 2007 problem and is returned to in Section 3.4). In the case that skills are learned in the workplace, it is obvious that there will be a better result when the teacher has large knowledge and experience than when he does not. But most of the technicians and engineers involved have not had any formal education in teaching. In the future, greater cooperation between companies and higher education is anticipated, to find ways to help the effective transfer of skills from masters to juniors. Specialist teaching groups, accumulating manufacturing know-how, are developing in technical colleges and universities for this purpose. An example of activities from such a group is given in Chapter 4.

Continuity of traditions of skill is vital to a company's maintaining a strong performance. There is a proverb that says experience and knowledge, once gained, remain for a long time. But it is not necessarily true. As an excellent worker becomes older, a company may choose to move him to a management function. This is particularly the case with those companies that operate a seniority-based personnel system in which a worker's pay may eventually become greater than his job can justify. It is important then that the worker can bring his master insights to his manager's job decision making. But over time, if nothing is done about it, those insights will begin to deteriorate. It is important to put in place measures for maintaining high skill levels in those who have moved to other responsibilities in a company, for example recognizing people as company treasures.

3.2 WORK DE-SKILLING

How to pass on technical skills, and the benefits that would arise if that could be done more quickly, started to be considered in the last section. The change in the nature of a job due to the introduction of new technology (CAD and also CAM—computer-aided manufacturing) has been mentioned as one way that has led to reduced learning times. Another way, in some ways similar, is to change the way in which the manufacturing process is organized. By splitting up a complicated task into a number of more simple ones, each part may be learned more quickly. This reduction of complicated tasks may be described as work de-skilling. Work de-skilling is nothing new. The next section gives some historical examples. But it has disadvantages as well as advantages, as introduced in the sections after that.

3.2.1 Historical Examples

The simplification of skillful work by replacing it by more simple, also mechanized, tasks can be traced back to the 1800s, when a production line for the manufacture of pulley blocks for sailing boats was built by Henry Maudslay (as described in L.T.C. Rolt, *Tools for the Job, A Short History of Machine Tools*, London: Batsford, 1965). By building separate sawing, boring, milling, and turning wood working machines and organizing them as a production and assembly line, Maudslay manufactured 160,000 ships' blocks in a year, employing only 10 men. Previously 110 men, each making all of a block, were needed. Another example comes from car manufacture. Before Henry Ford's time, cars were handmade by skilled carriage makers. Ford split the work into smaller parts, each of which could be learned more quickly. By this means and also by using conveyor belts to reduce heavy work and by buying some parts rather than making all of them, he created a system able to produce good quality cars more quickly. However, some dangers of de-skilling have been shown in Chaplin's film *Modern Times*, in which workers are treated as machines and lose interest in the quality of what they are doing.

3.2.2 Limits to De-Skilling

As introduced above, a production line sets up a chain of activities that ends with a product. Work at each stage is simplified. It is easier to learn. As a result, workers less generally skillful than before are able to carry out the work. The learning times become shorter too and productivity generally increases. However, product quality problems can arise.

A production line is a chain of activities in series. As already considered in Chapter 2, Section 2.3, the quality of the end product is limited by the quality of the weakest manufacturing link. As a product's manufacture is broken down into more and more small activities, the chance of one activity being a weak link increases. There is an optimum level of work division at which a company's performance will be best. This is illustrated in Figure 3.3. It shows how various measures of performance change with the number of stages in a manufacturing process. At first, as the number of stages (N) is increased from one or two to around five, there is a rapid increase of workers' efficiency at each stage (EW) at performing the simplified jobs. This outweighs any problems of increasing the chance of there being a weak link. An overall improvement in manufacturing performance

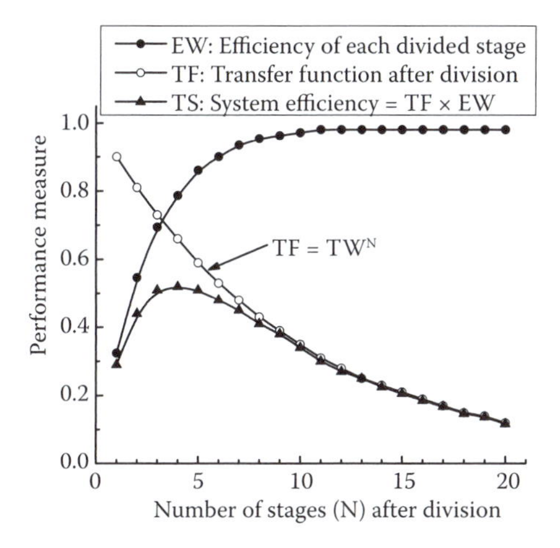

FIGURE 3.3
Effects of dividing work into N stages on the performance of each stage (EW), on the transfer function after division (TF), and on the performance of the system (TS). TF has been calculated for the case of the skill level of an expert worker: TW = 0.9.

is expected. However, as the number of stages continues to be increased, to more than 10 in Figure 3.3, no further improvements in stage efficiency can be gained. The chance of a weak link continues to grow. Even if there is no very weak link, but the skill level (TW) of each worker is less than 1.0, for example 0.9 in Figure 3.3, series connection causes a fall in the workers' combined level (as in Figure 2.10) to TW^N. TS, the system efficiency, which is the product of stage efficiency and combined level, falls too.

One can conclude from Figure 3.3 that when planning a manufacturing system, it is important to take into account what is an easy-to-reach skill level for the workforce and not to de-skill a job too much. There is a rational number of stages for a manufacturing system to perform at its best level. It will be one where workers find their tasks a little bit difficult but not too difficult to learn.

3.2.3 Mechanization and Automation of Skillful Work

Many manufacturing tasks have been mechanized or automated. More is expected. But the cost of automation, on the one hand, and the flexibility of human labor, on the other, will ensure that 100% mechanization and automation will never be desirable.

There are many cases nowadays where mechanization and automation have not been carried out, and people are still strongly involved in manufacture, even when mechanization and automation are entirely possible. For example, it might be supposed that manufacture of large numbers of parts, with little variety, is suitable for 100% mechanization and automation. In that case, the initial investment cost is shared between many parts, and, if there is little variety, the initial investment cost may not be so high anyway. However, even in that case an alternative is to break down the production process into a number of simple stages and use cheap human labor. Even if some of the tasks are mechanized and automated, dividing the work into stages and continuing to involve people gives flexibility and the ability more easily to change production to accommodate new products.

Developing this argument, the manufacture of simple parts with few stages between raw materials and the product is likely to be fully mechanized/automated. The manufacture of nuts and bolts or electrical resistors and capacitors and semiconductor memory are examples. On the other hand, a product with many stages in its manufacture is likely to involve human labor, particularly as it nears its final stages of manufacture and assembly, even if mechanization and automation are technically possible. In addition to the benefit of flexibility from human involvement that has already been mentioned, people are able quickly to detect and eliminate faults that may develop.

3.2.4 Skill Level and Automation

Many people now believe that automation of work in manufacturing requires just the pushing of buttons, and that engineers or technicians with conventional manufacturing skills are not needed anymore. However, an automated machine still needs to be operated by a worker. Considering what that involves enables what skills are still needed, and the importance of the machine's operator, to be properly understood.

Generally the smallest unit of machinery in a manufacturing company is the combination of machine and operator. The ability of both machine and operator influences the ability of this smallest unit. No matter how excellent is the machine's ability, if the operator's ability is low, the ability of the machine cannot be fully utilized. And even if the operator's ability is excellent, if the machine's ability is low, there is no way a fine product can be made. In the next paragraphs the differences between manually and automatically controlled machine tools are described, with an emphasis on what skills their operators need. Then a measure or index of

FIGURE 3.4
A manual lathe. (From Okuma Corporation, model LS. With permission.)

effectiveness of an NC machine and operator unit is developed, dependent on both machine ability and operator's skill. The skill level of the operator is still of great importance.

In the case of a manually controlled machine tool (as illustrated in Figure 3.4), material for a part (for example a steel bar of 100 mm diameter) is given to the operator. The part drawing, defining the part's dimensions, tolerances, and surface roughness, is also provided. The operator then takes the following actions:

- First, he plans, based on the material and drawing, how the work will be carried out. The plan includes determining such things as the machining methods, the order of operations, the cutting conditions (what cutting tool, spindle speeds, feed rate, depth of cut, cutting oil to use), and when to measure dimensions during the work.
- Then he prepares the tools needed according to the plan.
- He then loads the material on to the machine tool, sets the tooling, and starts machining.
- During machining, he measures dimensions (also as planned) to ensure they are within the drawing's range of tolerances.

In summary, with a manually controlled machine tool, only the material and the part drawing are passed to the operator. How to machine the part is left to the discretion of the operator in charge.

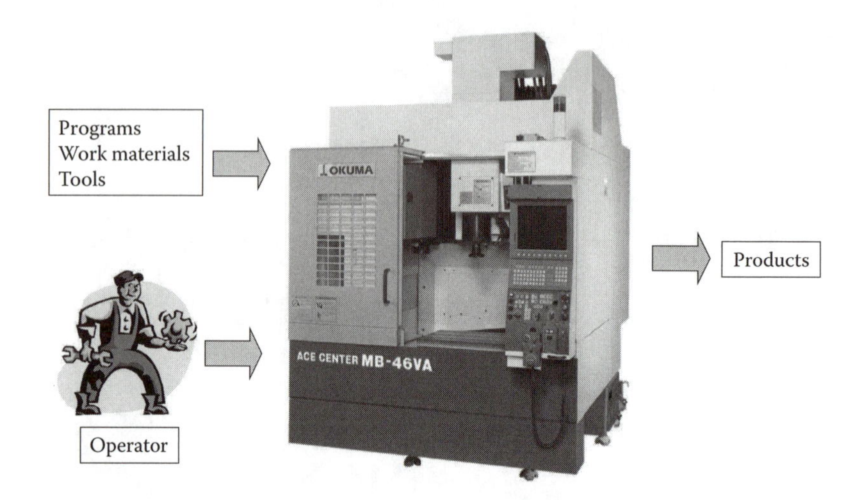

FIGURE 3.5
A precision machining center. (Machine tool from Okuma Corporation. With permission.)

An operator's responsibilities with an NC machine tool (as shown in Figure 3.5) are very different:

- The NC program needed to operate the NC machine tool is pre-prepared from the part drawing (nowadays from CAD data). This means the process planning (methods, order of operations, cutting conditions) is done in advance. The program includes all the procedures of machining. An accompanying machining specifications document defines how the work and tooling should be set in the machine. This preliminary work is usually done in the production engineering department.
- Based on the machining specifications document, the operator fixes material and tools in the machine and loads the NC program.
- Because the NC machine tool is controlled by the program, part measuring is not necessary while machining is going on.
- However, during processing, troubles such as vibration, tool breakage, and tool wear could occur. The operator must constantly monitor and take actions in response to such problems.

According to these activities, although the NC machine is highly automated, it is not 100% automatic. Someone is needed for preparing the NC data; the operator must fix materials in the machine, monitor the machine, and respond to problems. One way to respond is to change machine operating conditions such as spindle speed and feed rate. Operators are allowed to do that within limits. Spindle speed and feed rate are directly related to the rate of material removal; the higher the speed and feed, the higher the removal rate, other things being equal. In general, NC programs are written to keep the machining operation safe, i.e., so that trouble may be kept low while machining. This means they tend to be conservative with respect to choice of machining conditions such as speed and feed. However, increasing the spindle speed (cutting speed) and feed rate are effective ways to raise the process efficiency without changing the final form. Operators have responsibilities in this area too, balancing efficiency against increasing problems.

Thus, the effectiveness of an NC machine tool depends on the ability of its operator, even though some proportion of the tool's functions are automated. This qualitative statement may be made quantitative in terms of the learning curve, with the following assumptions/definitions, as shown in Figure 3.6:

- On a scale of 0 to 1, a particular machine tool may be considered to have a particular or specific effectiveness, E_E (for the same work content, machines can differ from one another in their effectiveness).
- A machine tool has a fraction R_A of its activity automated and a fraction R_{NA} that is not automated ($R_A + R_{NA} = 1$).
- The automated part is independent of the operator's ability.
- The nonautomated part depends on the operator's ability, E_W.

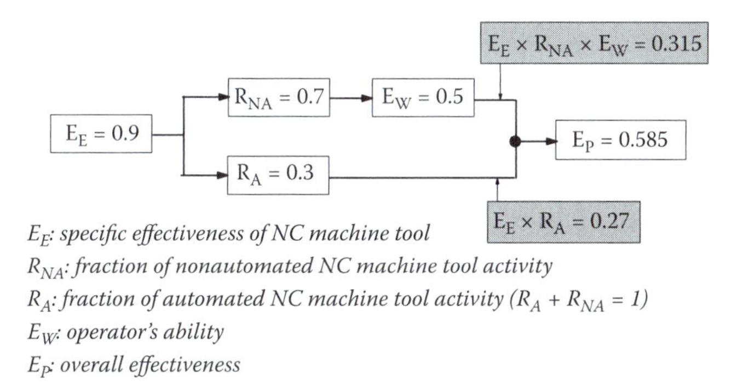

FIGURE 3.6
Influence of operator's ability on overall effectiveness of a numerically controlled (NC) machine tool.

Then, the overall effectiveness, E_P, of the machine and operator unit is

$$E_U = E_E \times (R_A + R_{NA} \times E_W) \quad (3.1)$$

For E_E and E_W both equal to 1, E_P naturally becomes 1 too. But consider the example in Figure 3.6. $E_E = 0.9$, $R_A = 0.3$ (so $R_{NA} = 0.7$), E_W (from the learning curve) = 0.5, and so $E_P = 0.9 \times (0.3 + 0.7 \times 0.5) = 0.585$.

According to machine tool manufacturers, $R_A = 0.3$ is a reasonable value for many NC machine tools. Even though those are commonly thought of as examples of advanced automation, their overall quality of performance is strongly influenced by the abilities of their operators.

3.3 THE SECURITY OF TECHNOLOGY TRANSFER

Up to this point, how to transfer technology and skill more effectively has been considered. However, some technology and skill may be confidential to a company. The problem then is how to stop it being transferred to other parties, either by accidental leakage or by intended actions, such as when a person takes his knowledge to another company on changing jobs, or even by industrial espionage.

As long as people are involved, there is a possibility of leakage no matter how strictly technology and skills are managed. Steps to prevent leakage are needed. Human resources, material things, and money are said to be the three elements that determine a company's success. Information should be added to this list, to make it four. Technology and skill are part of information. Steps to maintain security of technology and skill transfer may be summarized in terms of steps relating to human resources, material things, and information.

3.3.1 Human Resources

As people are involved in all the various forms of information, it is hard to prevent its leakage. There are even causes of leakage with which one can sympathize. For example, engineers and technicians may regret that their technology and skill would be forgotten if they did not pass it on somehow.

In thinking about how to prevent leakage of information by people, two situations can be imagined. One is leakage through people with authorized access to the information. The other is leakage by people who gain the information by accident. In the former case, leakage can be reduced by enforcing greater restrictions on who has access to the information. In the latter case, it should be ensured as much as possible that people have no motivation to pass on the information. These result in the following types of measures to reduce leakage:

- Clarification of which grades of worker can access the technical know-how and putting into practice systems to monitor access.
- Creation of personnel posts with the responsibility to improve the conditions of engineers or technicians in core technology areas.
- Building up an ethical culture while at the same time drawing up clear contractual terms relating to confidentiality.
- Maintaining contact with and supporting able and talented workers even after their retirement.
- Anticipating that people will change their jobs when designing information systems.

3.3.2 Material Things

The steps and equipment needed to make a product are the most important secrets of a manufacturing company. They are not revealed to those not directly involved, even though they may be workers in the company. To prevent the risk of leaks in this area, the following are commonly observed:

- What manufacturing processes are used to convert the raw materials to a final product are not disclosed to outsiders.
- The production process is divided up, and, wherever possible, each part is kept secret from the other.
- Those important parts of a product that depend for their manufacture on a company's original—core—technology are only made in a small number of places.

3.3.3 Information

With the exception of tacit knowledge, technological information like design data is normally held in records such as on paper or in computers. To prevent it leaking there are of course the people factors just considered. But also, it is possible to break down the information into separate parts and store them separately. Then only the parts that a person needs to know can be revealed to him, and it becomes difficult for a whole picture to be grasped. The division of the information can occur in a number of different ways, for example:

- Two types of design drawings can be made. One type is master drawings containing all the drawings' information. The other type is working drawings, which contain just the information needed for a particular job. Then only the working drawings are released to the workplace.
- In the case of manufacturing technology, the details of operation of an automated machine can be hidden even from the machine's operator.
- Work manuals can be written narrowly, one for each job, and supplied only to those who need them.
- Technological information can be classified as available only to workers above a certain rank. Which ranks can access what information can be made clear.
- It can be a policy not to document company special know-how accumulated over the years (although this may hinder the handing down of know-how).
- Important data can be kept only in paper, not digital, form, as digital data can be copied easily.

3.4 TURNOVER RATE AND TECHNOLOGY/SKILL TRANSFER

There are various data regarding the rate of worker turnover in Japanese industry. One source is the *Summary Results of Employment Trend Survey* published annually by the Japanese Ministry of Health, Labour and Welfare. The ministry defines the turnover rate as the number of staff leaving jobs in a year divided by the number of regularly (full-time) employed workers as of January 1 of that year, expressed as a percent.

Figure 3.7 shows the turnover rates in 2005 for a range of industries: manufacturing, IT, agricultural, finance and insurance, medical, and service industries. The turnover rate differs depending on industry. While service industry has a very high turnover rate of 22.3%, the turnover rate for manufacturing industry is approximately 11.7%, half of that of the

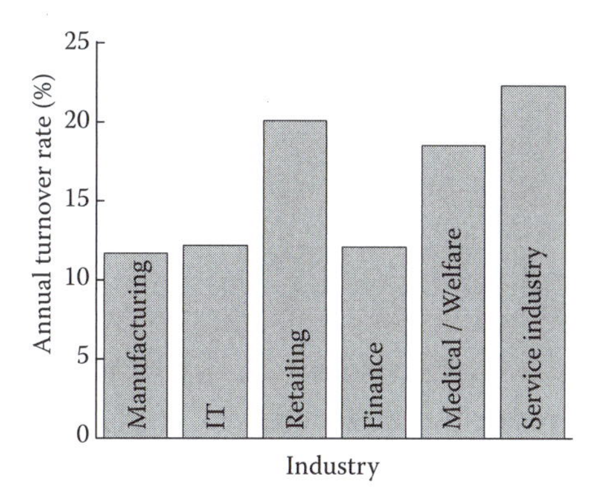

FIGURE 3.7
Annual turnover rates in 2005 for a range of industries. (From *Summary Results of 2005 Employment Trend Survey*, Japanese Ministry of Health, Labour and Welfare.)

service industry and the lowest of all industries in Japan. In addition, the turnover rate of all industries is 17.5%.

Perhaps the low turnover rate in manufacturing industry is because its work is particularly satisfying. For example:

- It is clear beforehand what is the work that needs to be done.
- The results of the work are easy to see and give a sense of achievement.
- A worker's improvement can be recognized quantitatively.
- It takes a long time to become a master; there is always room for improvement.

Particularly a sense of achievement, becoming a master, and room for improvement motivate job satisfaction and reduce turnover.

The 2005 survey has uncovered the following details of job turnover circumstances:

- Turnover rate is higher in the younger age group (35 years and below); especially 25.7% of those below 24 years old change their jobs.
- Turnover for management policy reasons increases for increasingly older age groups.
- For increasingly older age groups, turnover rate for personal reasons becomes lower. Instead, turnover occurs more for company reasons, for example expiration of the term of contract, management policy reasons, retirement age, etc.

In terms of workers' age groups, the above can be considered as reasons specific to younger workers, reasons to do with retirement, and reasons unrelated to age. From the point of view of technology and skill transfer, turnover can also be considered under three headings, almost the same as the age group reasons:

- Early turnover, within 3 years of employment
- Turnover of mid-level employee (service of 5 years or more)
- Turnover because of retirement age

The early turnover group is the one most strongly undergoing training. From a company's point of view the turnover problem is not the loss of skilled workers but the need to fill vacancies. Also, the continually large number of beginners causes the overall technology and skill level of the company to be lowered.

The turnover of mid-level employees is a more serious problem for a company than the turnover of beginners. Employees with more than 5 years of experience may be considered the backbone of a company. Depending on their job, they are starting to work independently and their technology and skills level is rising rapidly (as indicated in Figure 2.14). A company is beginning to recover its training costs with this group of employees. The group is the one, too, that is guiding beginners and from which veterans will emerge. Losing people from this group is a great loss to a company in terms of technology and skill transfer.

In the case of retirement, a company inevitably loses staff, and their skills, at the master level. It is a serious problem when a particular skill is indispensable to a company's activities. It is a growing problem because of the age composition in society. Figure 3.8 shows the number of Japanese people by age. The baby boom generation, born between 1947 and 1949, reached retirement age in 2007. This is known as the 2007 problem.

At the start of the 2007 problem, there was particularly a concern that there would be problems in the future with maintaining core computer systems such as the large general purpose ones used by companies to manage their financial affairs. These systems had been built, used, and maintained by the baby boom generation. Over the years they had been developed and modified as well. Companies are highly dependent on them. The systems are complicated structurally and difficult to handle for those who were not involved in creating them. It became a concern how such systems would continue and be upgraded and renewed once the baby

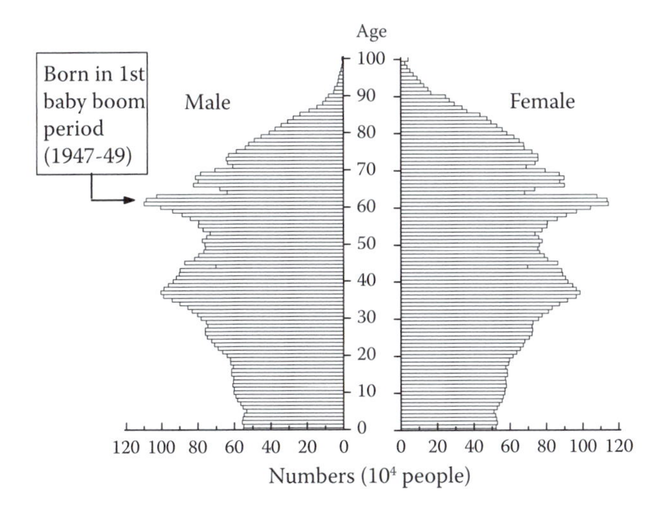

FIGURE 3.8
Age distribution of the Japanese population. (From 2010 census, Statistics Bureau of Japan, Japanese Ministry of Internal Affairs and Communications.)

boomers had retired. There are, in fact, many similar problems throughout industry, nothing to do with computer systems. What is more, such problems are nothing new, and how to transfer technology and skills is a question that has been much thought about. This has already been the subject of Section 3.1.

DISCUSSION QUESTIONS

1. Describe examples of explicit and implicit knowledge from as many circumstances from your experiences as you can (ten would be a good target).
2. For an activity or hobby that interests you, describe what you would consider to be skills at beginner, middle-ranking and expert levels. Imagine that you are in the position to advertise for a teacher at each of these levels: what would you include in the job descriptions for the posts?
3. "Reverse engineering" has not been mentioned in this book. Find out for yourself what is reverse engineering. Suggest how a manufacturer might guard his products and technology against unauthorized reverse engineering, with examples from products and technologies of your choice.

4

Virtual Manufacturing to Speed Up Learning

This chapter describes a particular manufacturing process known in Japanese as kisage work and in the West as hand scraping. It is a way of making very precise flat surfaces. It is extremely skillful work, and it is a very difficult technology to transfer, including to hand down the required skill to successors. It requires not only a high practical level of manual skill but also good judgment. Particularly the good judgment is difficult to transfer.

Although there are many skillful tasks in manufacturing, hand scraping is chosen here as a concrete example because it is particularly important. Examples of surfaces finished by this method include machine tool slide ways and surface plates of form-measuring machines. It is the flatness of the surface plates that determine the accuracy of the measuring machines. Hand scraping enables such slide ways and surface plates to be made more accurately than they could be made if the machine tools themselves were used to make them. As a special feature, the flat surfaces made by hand scraping are covered by fine dimples that act to trap oil. In the case of slide ways, this enables them to slide more smoothly.

Scraping is also chosen because how hand scraping is carried out by skillful workers has been the subject of study. As a result, it has been possible to create a computer-based simulator of the work. It can be used as a learning aid, to reduce the learning curve time for the work, in the same way as a driving simulator might be used to increase the skills and judgments of a learner driver or a flight simulator may be used by a trainee pilot. The development of such aids may have great benefits in the future as a means of improving and speeding up technology transfer.

The next sections describe hand scraping and a study of the skills and strategies developed by master workers to carry it out. How the process understanding from the study has been used to develop a computer simulator is then described. Finally, the simulator's uses and benefits in the education of engineers are discussed.

4.1 HAND SCRAPING

Hand scraping is a process that enables a high-precision flat surface to be created even when a standard reference flat does not first exist. That is why it is so important, even today. The starting point is three nominally flat surfaces, i.e., surfaces that are meant to be flat but may have some unintended curvature or high spots on them. Suppose that they are arbitrarily labeled A, B, and C. The worker first takes A and B and, using A as a reference, removes the high points on B relative to A. He then takes B as a reference and removes the high points on C relative to B. He then takes C as a reference and removes the high points on A relative to C. This cycle is continued. Gradually all three surfaces are brought closer to flatness, until some required standard is met and the surfaces could themselves be regarded as reference surfaces. Once a reference exists, then other surfaces can be scraped to match it. The removal process is by hand, using a tool like a chisel, which in one action may remove material only a few microns deep or even less. The judgmental skill that is required is deciding from where on a surface to scrape away the material. The manual skill is in controlling the amount scraped away in one action. It is not possible to replace material if too much is scraped away.

The flow of activity in judging from where to remove material, and removing it, for one pair of surfaces, is shown in Figure 4.1. First, there is marking a surface to help identify where are the high points. Then there is interpreting the marking (judgment), to recognize from where to remove material. Then there is the scraping work, before remarking the surface to repeat the cycle until done.

Figure 4.2 gives details of the marking process. There are two ways to do this. In both, one surface is identified as the standard or reference surface and the other as the surface (or workpiece) from which material is to be removed. At an early stage of scraping, when there may be large deviations from flatness, the standard surface is coated with a thin layer of paint. For

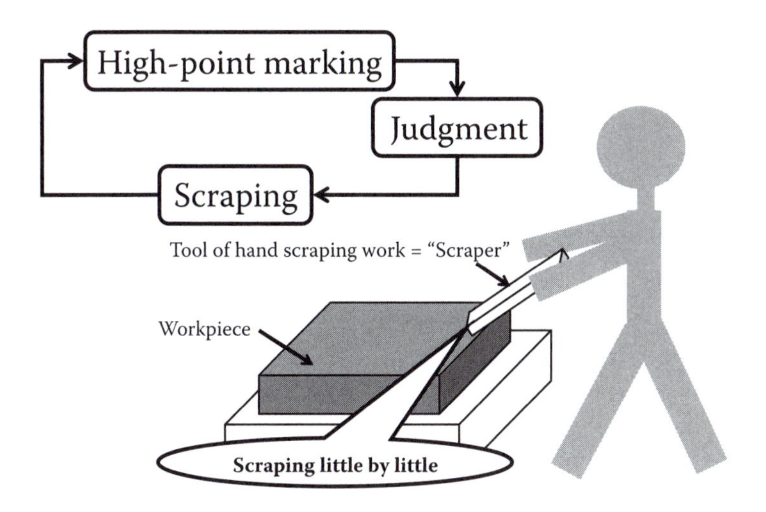

FIGURE 4.1
Hand scraping.

example, red lead primer is commonly used, which is a vermillion color. The standard surface is then rubbed against the work surface. The paint transfers to the high spots on the work. This way of marking is known as positive transfer because paint transfers from the standard to the work surface.

At a later stage, when the surfaces are more precisely flat, or if the starting surfaces have perhaps been preprocessed, for example, by surface grinding, it becomes more effective to paint the work surface. Then, when

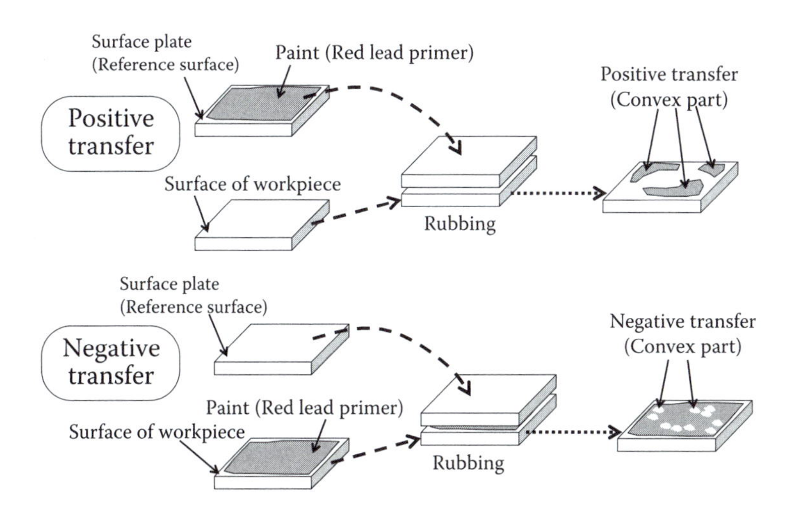

FIGURE 4.2
High-point marking.

it is rubbed against the standard surface, paint is removed from the high spots. This is known as negative transfer.

As already mentioned, the worker's skill to scrape precisely the right amount needed is very important. This skill can be learned through practice and training. How to judge where to scrape is equally important. It is this that is gained through experience and is a know-how that is difficult to teach. How an experienced worker makes judgments is the topic of the next section.

4.2 AN EXPERIMENTAL STUDY OF EXPERT SCRAPING JUDGMENTS

How a skilled worker scraped a cast iron plate of size 240 × 230 mm (which is a standard surface plate size) has been followed. The negative transfer recognition phase of the process was chosen for the study, as it is closer to the finishing stage than is the positive transfer phase. The worker's progress was followed over several cycles of marking, recognition, and scraping. It was necessary to find a way both to measure how the plate's surface form was altered from cycle to cycle as well as record the patterns of paint transfer on which the worker was basing his judgments. This had to be done without interfering with the worker's activity.

The method chosen was for the plate to be scraped while mounted on the bed of a numerically controlled (NC) milling machine. A laser scanning surface displacement meter was mounted on the spindle nose of the milling machine. It was able quickly and automatically to scan over the plate to record the plate's 3D form. The 3D form was then digitized and sent to software that created a 3D relief map of the surface on a computer screen. A CCD camera was also used to photograph the patterns of paint transfer from the work surface. As the paint coating was very slight, image processing was needed to enhance the contrast between coated and uncoated areas.

It was not practical to record form and coating changes while the worker was scraping. Not only would the amount of data gathered have been huge, but it would have interfered with the work. Instead, recording was undertaken at the start and end of a cycle. As a result, only a partial history of the work's progress was recorded. Some details of the worker's strategy were not captured. However, an outline of strategy was obtained.

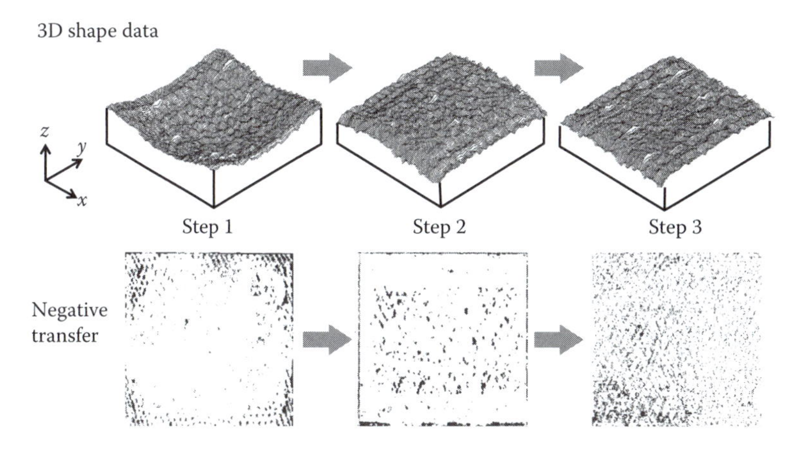

FIGURE 4.3
A series of hand scraping steps by a skillful worker. In the lower row dark areas are regions from which paint has been removed. (From T. Sugino et al., *Journal of the Japan Society for Precision Engineering*, 69(7): 949–954, 2003, in Japanese. With permission.)

Figure 4.3 shows progress of scraping from the start (step 1), to an intermediate stage (step 2), to completion (step 3). At the start, in this case, the work surface was concave, with four raised corners. The relief map showed that the corners were raised about 17 μm above the level of the center. However, all that can be deduced from the CCD picture is that the corners are high. Obviously, anyone could deduce that. It is further judgment relating to how high are the corners that is part of a skillful worker's experience.

After several cycles of marking, judging, and scraping, step 2 was reached. In contrast to step 1, the CCD view shows that paint has now been transferred from the center of the work surface. The surface has been changed to convex. At first it looks as if too much material has been removed from the corners. In fact, it has been done on purpose. This is considered further in Section 4.3. The worker's experience and mastery of skill is probably demonstrated in this apparent overshoot of material removal.

Step 3 shows the finished plate. Separate surface form measurements confirmed that the plate was flat to within the required JIS (Japan Industrial Standard Association) standard. The relief map shows roughness resulting from scraping marks that might cause people to think the surface is not flat. However, the even distribution of negative transfer paint marks seen in the CCD view confirms there are no large areas of low or high height, i.e., that the surface form is flat.

As illustrated above, the worker's information about the surface form during scraping is limited to the visual information from paint transfer to or from the work surface. How much can be judged from that about the form of the surface that has not yet been revealed is part of the worker's skill. Although at step 1 there are large areas at the corners from which paint has been transferred that make it easy to judge that the high points are at the corners, there are also other areas where there has been less transfer and which can therefore be judged to be almost as high as the corners. How accurately a worker can judge this kind of situation is a key skill in scraping. If there is a mistake in judgment and too much is scraped away, it is very difficult to correct it.

It is possible to extract some more detailed information from the experiment. Between steps 1 and 2 in Figure 4.3, at least a depth of 21 μm was removed from the surface, according to the relief map information. Only 3 to 5 μm is removed between marking stages. It can therefore be concluded that going from stage 1 to 2 took at least four cycles of marking, judging, and scraping.

From all this, the importance of judgment and strategy in scraping can begin to be seen. Also, it is clear that the manual skill of achieving the amount of material removal that is intended is also important. The fact that there are two things to be mastered at the same time (judgment and manual skills) makes mastering scraping skills even more difficult. Without manual skill, there is little opportunity to develop judgmental skill. Without judgmental skill, what manual skill should be aiming at is unclear. There would be great benefit if the two aspects of learning could be separated. Another problem in learning scraping skills is the long time that it takes to complete a real scraping process. Several days is common. Learning by practice takes a very long time and is very expensive. For all these reasons, i.e., separating the different skill aspects and possibly shortening the on-the-job practice needs, a virtual scraping simulator would be a benefit. This is returned to in Section 4.4, after considering in more detail hand scraping strategies that are developed by master workers.

4.3 HAND SCRAPING STRATEGY

This section builds on the observation that a master removes more material between steps 1 and 2 of Figure 4.3 than is apparently necessary to transform the work surface from a concave to a convex form.

There are two goals of scraping. One is to achieve the required flatness accuracy. The other is to achieve that in the shortest possible time. To achieve scraping in the shortest possible time means to create a flat surface with the smallest number of marking, judging, and scraping cycles. Given that the depth removed per cycle is approximately fixed by the scraping tool, to create the flat surface with the smallest number of scraping cycles means to create a flat surface with the smallest volume of material removal. This implies that the worker must quickly form in his mind some image of the overall form of the surface and what the new surface form would look like after scraping, rather than just have a view of individual, unrelated, high spots.

An overall view of the form could be completely obtained using a 3D measuring machine. The question is how can such an overall view quickly be inferred from the partial knowledge gained by rubbing two surfaces together and noting from where paint is transferred from one to the other? A possible answer comes from the step 2 condition in Figure 4.3. After scraping the edges so that they become lower than the center, the next cycle of surface painting and negative transfer from high points allows the previously hidden form detail toward the center to be made visible. This can be generalized to a statement of strategy. Initially, removing a little more material from the highest points than apparently is necessary allows the form of a greater area of lower parts to be revealed in the next cycle of marking. This scraping of apparently too much material should continue until a complete overview of the surface form is obtained.

A scraping strategy based on this, for the example of Figure 4.3, is shown schematically in Figure 4.4. In this figure the dark areas are areas higher than their surroundings. It is supposed that the maximum depth scraped away in one cycle of activity is d μm, and the number of cycles from step 1 to step 2 is n_1, and from step 2 to step 3 is n_2. Then the maximum depth S_1 removed between steps 1 and 2 is $d \times n_1$, and the maximum depth S_2 between steps 2 and 3 is $d \times n_2$. It is further supposed that the transfer of paint caused by rubbing occurs over a depth range of highest points that is the depth marked "a" in Figure 4.4. This creates an uncertainty in estimating the high-point boundaries. The range arises because the reference surface plate is not completely smooth. It has unevenness as a result of its own scraping marks. The rubbing action itself also contributes to the depth range.

Then, from step 1 to step 2, the material from which paint is transferred is scraped away by the relatively large amount S_1, until negative transfer

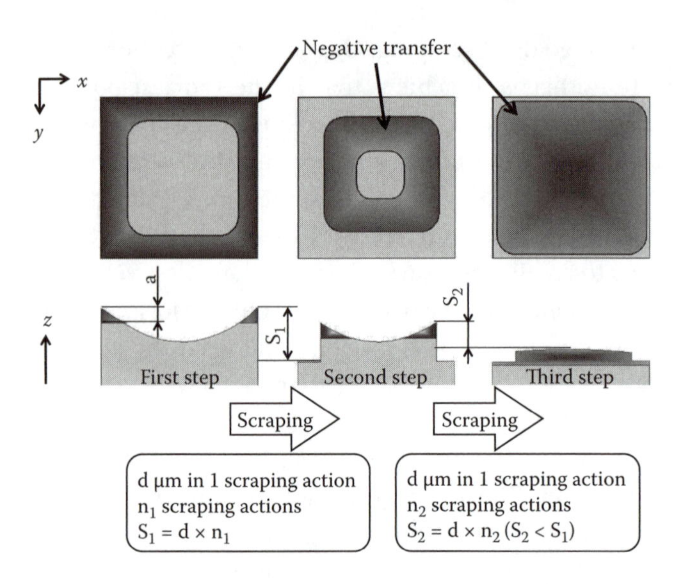

FIGURE 4.4
A hand scraping strategy, based on Figure 4.3's negative transfer experimental study. The strategy is equally applicable in the case of positive transfer. (From T. Sugino et al., *Journal of the Japan Society for Precision Engineering*, 69(7): 949–954, 2003, in Japanese. With permission.)

and judgment of what material to remove extends right to the center of the plate's surface. The surface at depth S_1 that then exists as a rim round the edge of the plate defines a depth that scraping from step 2 to step 3 should not exceed. Once step 2 is reached, progression to step 3, i.e., finishing the process, can proceed effectively. It is important that S_2 does not exceed S_1. A way to ensure this is necessary. It can come from noting the increase in negative transfer area from one marking stage to the next, equivalent to estimating surface slopes on the plate, and hence anticipating what depth removal will complete the scraping process. This ability to estimate surface slopes is a major part of the skill of master scrapers.

Figure 4.4 is just one example. For different examples, i.e., different surface forms at step 1, the strategy of apparently removing too much material until step 2 is reached will take different geometric forms. However, one principle is common to all cases. The worker should as quickly as possible be seeking to gain an overview of the surface form to be corrected.

Figure 4.4, based on Figure 4.3's experimental study, is for the case of negative transfer. However, it is equally applicable in the case of positive transfer.

4.4 COMPUTER SIMULATION OF SCRAPING

As previously mentioned, strategy is just one part of scraping skills. Manual skills are also important but take a long time to develop. Mistakes made during real scraping are not easily put right. Once too much material is removed, it cannot be replaced. Learning by practice is time-consuming and can be dispiriting. Trying to learn strategy by observing how a skilled worker carries out the real process is equally difficult. It is for all these reasons that a virtual simulator of scraping would be useful. It would allow mistakes to be made and undone. It would allow different strategies to be explored and discarded, all much more quickly than in real life. This section describes such a computer-based simulator, with an example of its operation.

The structure of the simulator is shown in Figure 4.5. It contains three modules: high-point marking, judgment, and scraping. They are connected in a cycle in the same way as the real work of scraping (Figure 4.1). Each module is described as follows.

4.4.1 High-Point Marking

First, the 3D surface relief data from either a real work surface (for example, from Figure 4.3) or an artificially generated surface are entered

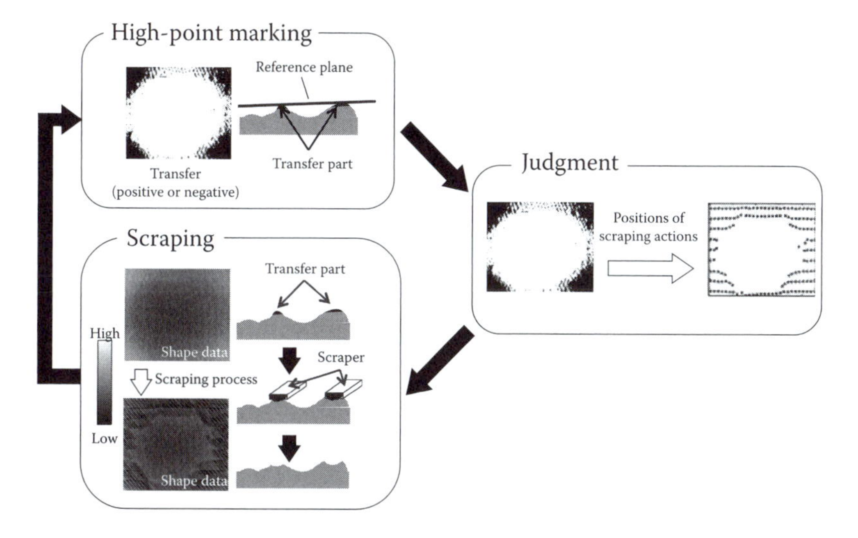

FIGURE 4.5
Outline of computer simulation of scraping. (From T. Sugino et al., *Journal of the Japan Society for Precision Engineering*, 69(8): 1104–1108, 2003, in Japanese. With permission.)

into the system. In one mode of operation, the three highest points on the surface are identified and the plane through those points is defined. This is equivalent to a perfect reference surface. Then everywhere where the gap between the reference plane and the work surface is less than the value a (Figure 4.4) is determined. These are the high points. During the marking process it is to these areas that paint is transferred (positive transfer) or from which it is removed (negative transfer). The module has a graphical user interface. It shows the pattern of transfer areas. The module can also be used to simulate the three surfaces' version of scraping (surfaces A, B, C of Section 4.1). Instead of entering data for one surface and comparing surface heights to a reference plane, data are entered for two work surfaces. They are placed face-to-face and moved over each other to determine areas of contact and transfer.

Figure 4.6 compares an example from the simulator with a real-life case. It is for negative transfer marking from a cast iron plate 200 × 200 mm in size, placed against a reference surface. Although the simulated transfer areas show less fine detail than do the real ones, the coarse details are the same.

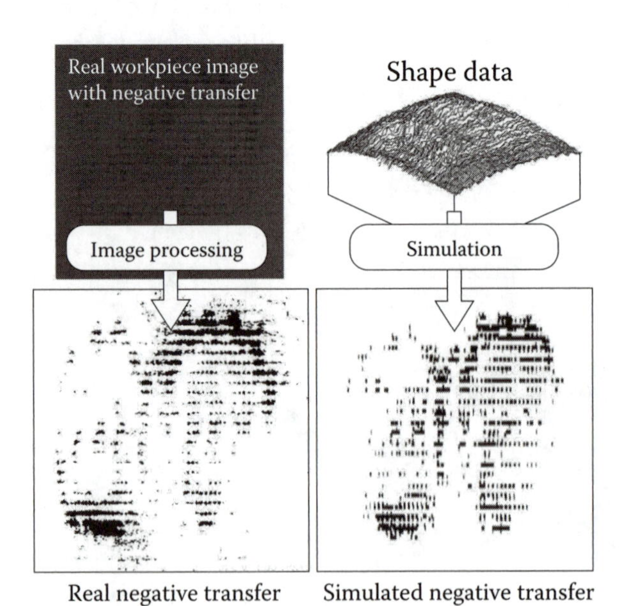

FIGURE 4.6
Real negative transfer compared to simulated negative transfer.

4.4.2 Interpretation and Judgment

This module can be used in either of two modes. One mode is automatic. The computer interprets the transfer patterns and decides from where material should be scraped. The other mode is manual. A person decides from where material should be scraped. The manual mode is the subject of Section 4.5. Here the automatic mode is described.

In automatic mode a set of rules is used for decision making. A rule consists of a recognition part and an action part. Two examples of simple rules are given here. The recognition part is the same for both, but the action parts are different.

The recognition part classifies separate high spot areas according to whether their areas are smaller or larger than the expected size of a single scraping mark. If an area is smaller than a single scraping mark, the position of its center of area is calculated and labeled. If an area is larger than a single scraping mark, the area is segmented into an array of touching scraping marks. The positions of the centers of area of each segment are calculated. The left-hand part of Figure 4.7 shows an example of both types of area, the actions of classifying them being described as classification by labeling and classification by segmentation.

The simplest action after recognition is to remove a volume equal to that of a single scraping mark from each of the centers of area. An example is shown in the right-hand part of Figure 4.7. This is a simple action that a beginner might take. A more complicated action is also to remove material

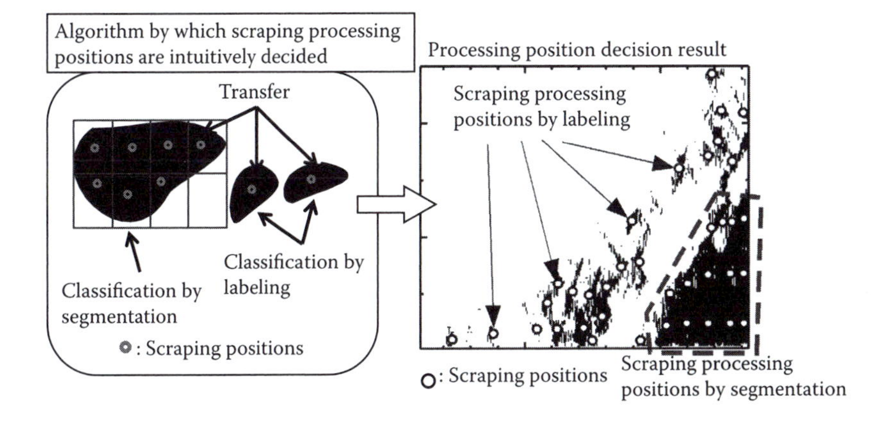

FIGURE 4.7
An example of transfer area recognition and subsequent processing position decision making.

from between two centers of area if their separation is less than some value. This is an action that might come from experience.

Then the two rules are:

Rule 1. Remove material equal in volume to one scraping action from each of the identified scraping positions.

Rule 2. Remove material as in rule 1 and additionally remove material from between identified scraping positions if these are closer than some set value.

An example of the application of these rules is given in the next section.

4.4.3 Scraping

In this module the surface form is altered according to the decisions from Section 4.4.2. How this is done is shown schematically in the upper part of Figure 4.8. The module has in its database a 3D unit shape that represents one scraping action. This can be subtracted from the surface form, as many times and as at many positions as are demanded. The lower part of Figure 4.8 gives an example. In fact, there is a library of unit shapes that simulate different single scraping actions. They depend on the shape of the scraping tool. Also, shapes made by skillful workers can be measured,

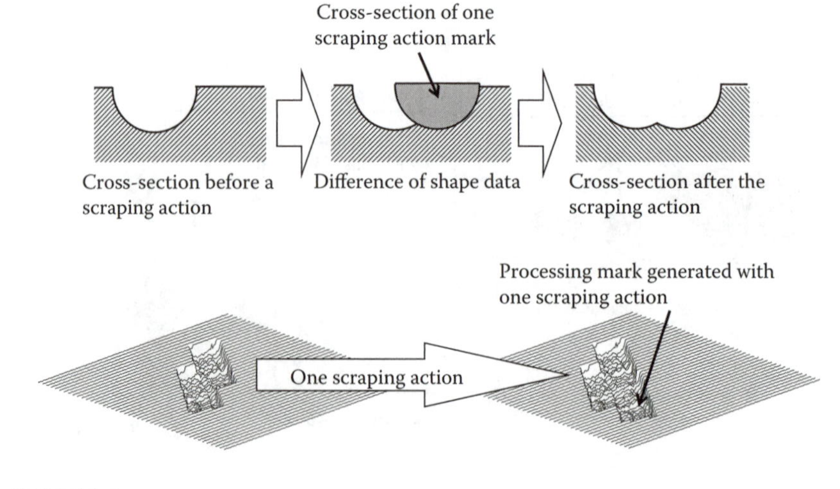

FIGURE 4.8
Reproduction of a scraping action by differencing between workpiece and processing mark.

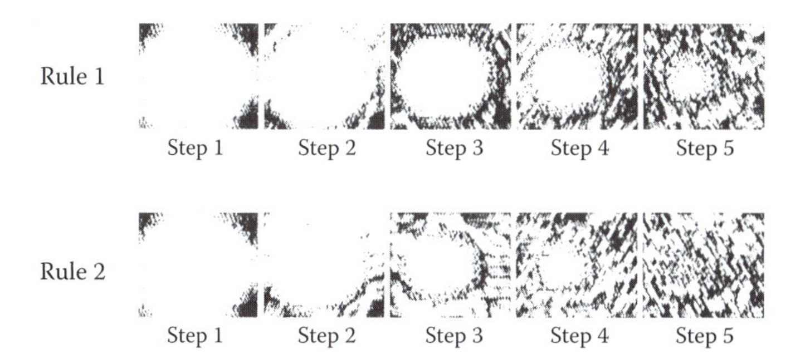

FIGURE 4.9
Examples of computer simulation with two different rules.

digitized, and stored. In this way, the actual appearance of a scraped surface can be reproduced.

Figure 4.9 shows the progress of scraping for the example from Figure 4.3, when rules 1 and 2 from Section 4.4.2 are separately applied. Applying rule 2 results in complete scraping of the surface after only five cycles. An unprocessed area still exists at the center after five cycles of applying rule 1. In this example, use has not been made of the strategic thinking of Section 4.3. Even so, the example shows that one strategy can be better than another. It demonstrates how simulation can be useful. A range of strategies can be explored as part of, in this case, the scraping learning process.

4.5 COMPUTER SIMULATION AND EDUCATION

The complete skill of scraping requires manual skill to be coordinated with judgment. Although there is no substitute for real scraping for developing the manual skill, at least learning the judgmental part may be speeded up by simulation. Then the confidence concerning that part may increase motivation for learning the manual part.

Figure 4.10 shows use of the simulator in its human-controlled decision mode, as it is used for education. The judgment part of the cycle is operated manually by the learner. As has been written before, it allows people to undo and redo actions, to test different strategies. Also, by changing the initial form of the surface, a large number of practice scenarios can be created. Experience can be built up in this virtual way.

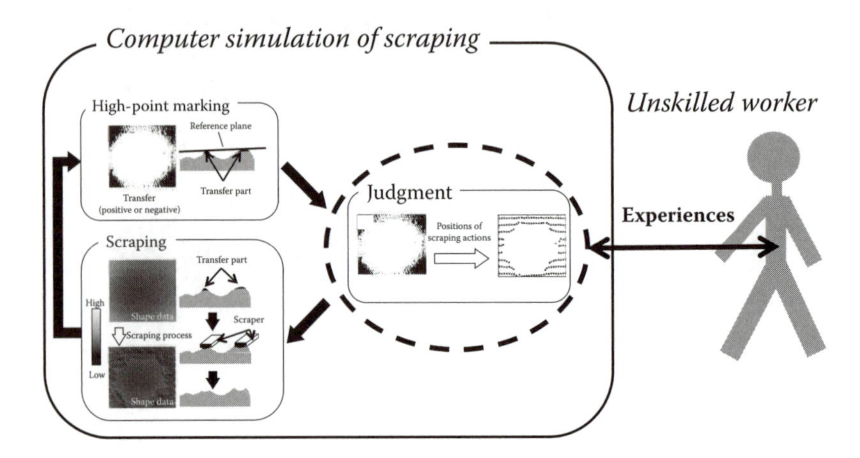

FIGURE 4.10
Computer simulation and education.

The user interface is particularly important, in order to hold the user's attention and stimulate his imagination. A haptic interface, some sort of joystick that feels like a scraping tool, would be ideal. In that case, of course, it might even be possible to teach the manual skill of scraping by virtual means. But that does not exist at present. Instead, clicking the computer mouse is used to generate scraping marks and change the surface form (as shown in Figure 4.8). It creates a sense close to reality.

However, having to click the mouse too many times, because only a small amount of material is removed each time, can cause a learner to become uninterested. When a large number of scrapes are required in a single area, for example in the case of the large island (processing by segmentation) in Figure 4.7, it becomes advantageous to be able to remove all material with a single click. For this reason, the simulator has a lasso tool, as used in photo-editing software. It enables the user to define an area to be altered collectively. The working of this user-controlled option is shown in Figure 4.11. It allows scenarios to be explored more quickly.

The simulator can also be used to help a learner start to visualize the 3D form of a surface from the 2D paint transfer areas that are the only information available in real scraping. As was written at the end of Section 4.3, it is this that is the strategic skill of an expert scraper. The 3D surface relief data input to the high-point marking module of the simulator (Section 4.4.1) can be displayed as a map alongside the 2D transfer area image, as shown in Figure 4.12. By showing a learner many different cases from real life, a virtual experience can be built up.

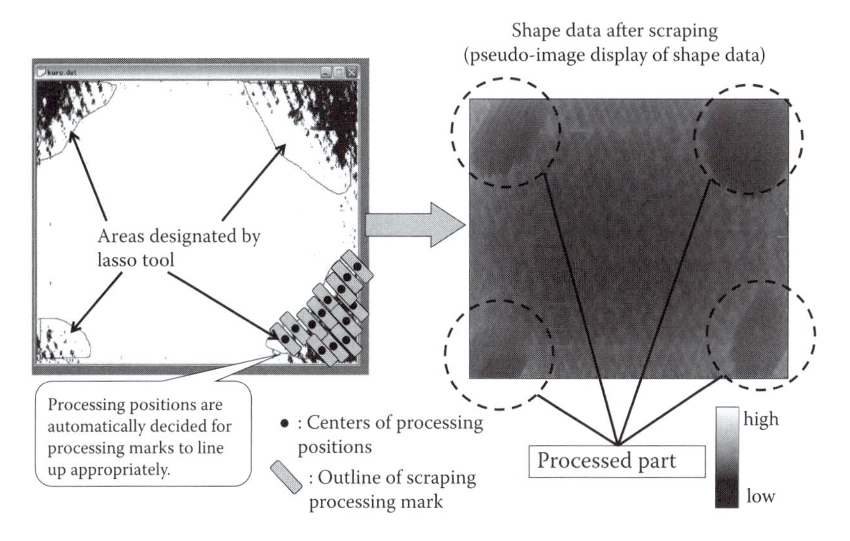

FIGURE 4.11
Hand scraping processing of area designated by lasso tool.

Currently manufacturing education does not generally make use of simulators. It is easy to imagine from the example here that a development of simulators would shorten learning times and hence quicken workforce training. It would make technology transfer easier. Of course, some real experience will always be needed. But experience from, for example, car driving simulators suggests that large benefits would result.

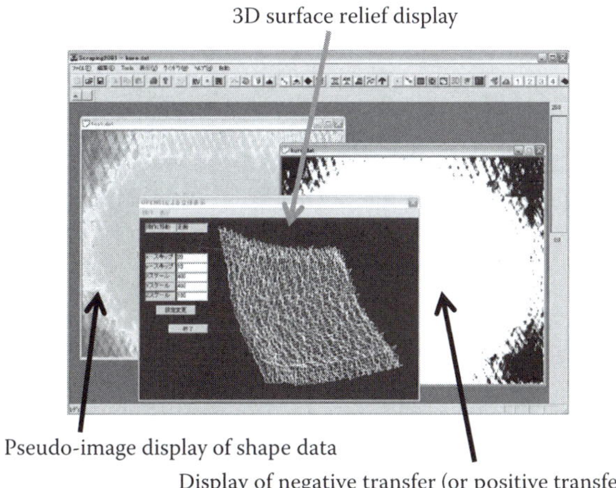

FIGURE 4.12
Screenshot of scraping simulator.

DISCUSSION QUESTIONS

1. Investigate some example of skillful work requiring technology transfer in manufacturing industries.
2. For the example of question 1, analyze the structure of the skillful work and discuss the reproduction of the skillful work with a computer in the same way as in this chapter.
3. Discuss the key points of technology transfer of skillful work in manufacturing industry based on the contents of this chapter and the above two questions. Moreover, discuss the points expected to be difficult about technology transfer of skillful work.

5

Production Management and Technology Transfer in Manufacturing

Technology transfer involves transfer of management technologies as well as product and process technologies. What management technologies are important depend on the surrounding business and society conditions. There is a wide range of business and society conditions in which companies make decisions about technology transfer. This chapter is concerned with these broader issues of technology transfer. Section 5.1 is particularly concerned with what is the scope of production management. Section 5.2 introduces the idea of the product life cycle and what management technologies and manufacturing strategies are most appropriate at different stages of the cycle. Technology transfer between companies and countries is a way to make the best and most profitable use of finite resources. It depends on the state of development of both companies and their strategic views of their places in the market. What is appropriate technology transfer is the subject of Section 5.3.

Understanding the breadth of factors, concerns, and reasons behind decisions to transfer technology helps to understand why technology transfer takes so many forms. This chapter, as well as having a value of its own, may be considered introductory material to Chapters 6 to 8.

5.1 PRODUCTION MANAGEMENT

In this first section the activities of production are described from a production management point of view. Section 5.1.1 explains the breadth and scope of production management and the broad demands placed on it.

How to satisfy the demands has different constraints and takes different forms, depending on what are the products being manufactured and also what are the products' quantity and variety. Further details on this topic are given in Section 5.1.2.

5.1.1 Production Activities and Management

The activities of introducing a new product to the market need different technologies, depending on the product's stage of development. It is common to divide development into four stages, as shown in Figure 5.1. These are research and development (R&D), product development, process development, and finally production itself. The different technologies at the different stages are listed in Figure 5.1, from core technology to production management technology.

In Figure 5.1 the linking of production management technology to production reflects a narrow definition of production management. Production activities depend on the product and process development that have preceded production. The product and process should have been developed with production and its management in mind. Thus, in a wider sense, production management is involved not only with production itself, but also with product and process development. Production management technologies are developed to manage product and process development as well as for managing production itself.

From a business point of view, production in an organization can be considered as conversion and value-adding activities. As shown in Figure 5.2, these start from financing and continue to procurement, production, and sales. From this point of view, technologies for planning and controlling production are only part of production management. It is necessary for

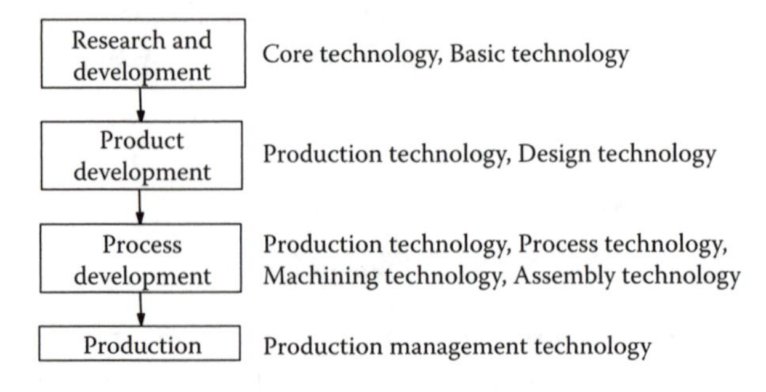

FIGURE 5.1
Activities and their technologies.

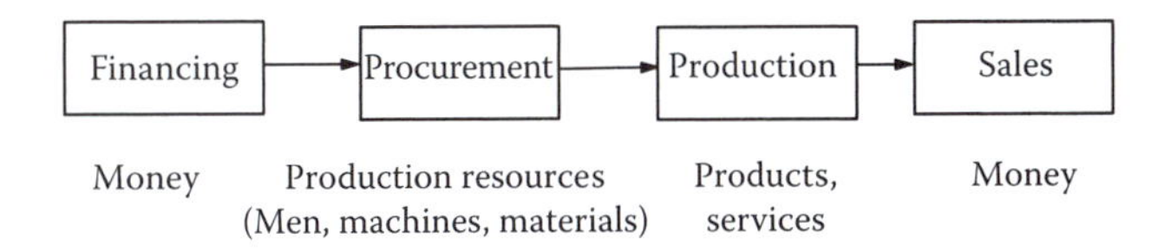

FIGURE 5.2
Value-added activities in an organization.

production management also to include financing activities, procurement activities, and sales activities.

Within the general activities of management, the purpose of production management is to plan and control production activities to produce products and services economically, with the required quality, at the required quantity, and when needed. With that definition of purpose, the goals of production management can be summarized under the headings of (1) specification and quality, (2) quantity and time, and (3) price. Under these goals, production management must accomplish the following:

- Guarantee the specification and quality of products and services as demanded by the market and by society.
- Satisfy the volume and time demands from the market and society, taking into account production lead times.
- Satisfy the specification and quality of products and services and the volume and time of demands efficiently so that the prices can be kept low.

Production management responds to these three elements of demand by managing the production process, which may further be considered also to involve three elements: humans (workers), machines, and materials. In terms of these three elements of demand and three elements of process, production management may be regarded as made up of a range of management technologies. Quality management, quantity and delivery management, and cost management address the three elements of demand. They may be considered the primary production management technologies. Personnel management, facilities management, and materials management address the three elements of process. They can be regarded as the secondary production management technologies. Furthermore, all these must interact with the managements that surround them.

5.1.2 Production Systems and Their Features

Generally, as shown in Figure 5.3, production systems are one of three types, suitable for different demands of product variety and volume. They are jobbing (or job shop) production, suitable for products with a large variety and small numbers of each product; lot (or batch) production, for smaller variety but larger numbers; and line production, with least variety and largest numbers of products.

In the case of jobbing production, the repeat demand for any particular product is small. A general purpose production facility is prepared. Machine tools are laid out in the factory to respond equally to the demands of any product. At the opposite extreme, in line production, a production facility is arranged as a linear series of stations for the exclusive continuous production of one product. In between, lot production fulfills the need for medium variety and volume demands. Machines are organized as a number of general purpose facilities, each of which can be used for a special purpose by changing setups. This organization increases productivity.

In the case of job shop production, products are made individually to a customer's specification. The management question is how to produce the orders quickly and efficiently. By contrast, for continuous line production the question is how to respond economically to varying demand for a product while at the same time keeping the production line running. For lot production, it is expected that different products will be manufactured at different times. Effective management is concerned with what is the best way to respond to changing customer orders. Questions to answer are

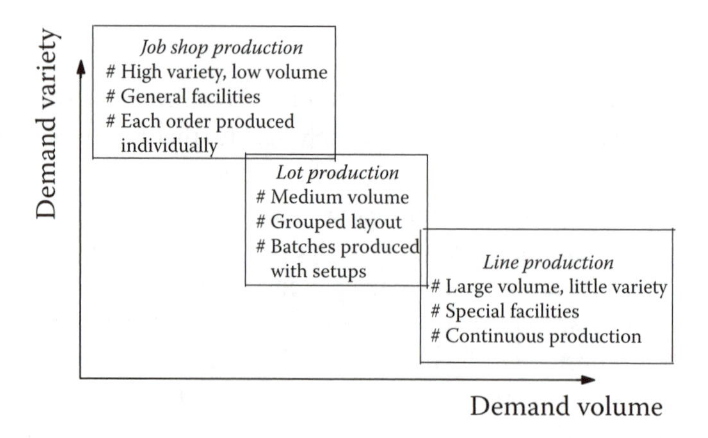

FIGURE 5.3
Process structures based on demand variety and demand volume.

what is the required lot size and how to inject new orders into the manufacturing schedule. It leads to promoting commonality and standardization of parts between products that are made in lot quantities.

5.2 THE PRODUCT LIFE CYCLE

The range of management technologies and production systems is described in the last section. It is written that production management technologies are developed to manage product and process development as well as for managing production itself. It could have been added that within production the appropriate management technologies change with time, from the introduction of the product, through a period of growing sales, and eventually to the time at which the product begins to be displaced by other products. The cycle from birth to death of a product is called the product life cycle. This section considers management technologies and production systems in the context of the product life cycle, especially which management technologies are most important and which production strategy is most appropriate at each stage of the product life cycle.

5.2.1 Management Technologies in the Product Life Cycle

Management throughout a product's life cycle aims to satisfy the different demands of the different stages. Figure 5.4 (developed from R.H. Hayes and S.C. Wheelwright, *Restoring Our Competitive Edge: Competing through Manufacturing*, New York: John Wiley, 1984, with additional material) shows four stages. In an introductory (or start-up) period a new product is developed and introduced to the market and is evaluated by customers. If it satisfies a demand, it enters a period of rapid growth. If the product becomes an established success, demand for it will reach new heights in a period of maturity. Later the product will either enter a period of decline in which it is replaced by new products or, in some cases, will become a commodity in which it survives in the market in a state of steady demand (this is the case shown in the figure).

The product variety, demand, industry structure, and form of competition all depend on the life cycle period in which a product finds itself. The text within Figure 5.4 describes these.

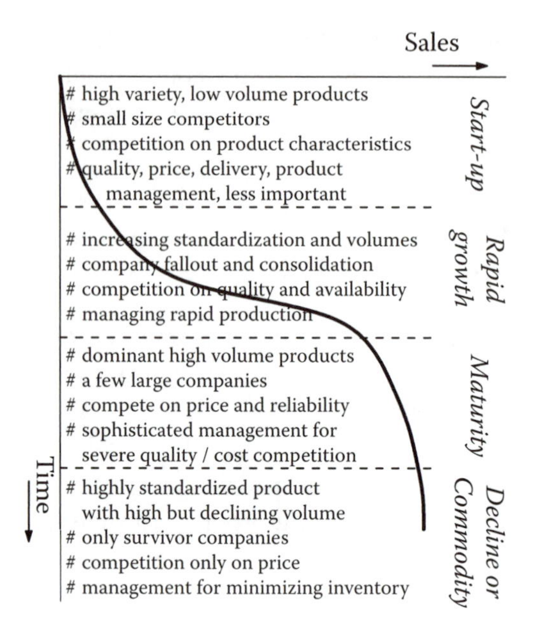

FIGURE 5.4
Product life cycle and management technologies. (Developed from Hayes and Wheelwright, *Restoring Our Competitive Edge: Competing through Manufacturing*, New York: John Wiley, 1984, with additional material.)

In the start-up period, the market consists of a wide variety of products with low demand. Many small companies make up the industry, competing with each other for market share through special features of their products. In the rapid growth period, as the standardization of product type progresses, the variety of products decreases and the demand for each type of product increases. The number of surviving companies reduces and each gets larger. Competition becomes based on product quality and performance.

In the maturity period, a dominant product type emerges. High-volume demand arises for a limited variety of products. This leads in most cases to a few dominant companies. At this stage, competition is based on reliability and product price. In the last declining or commodity period, either demand decreases little by little or a great demand continues for a highly standardized product. The remaining industry consists only of companies that survived the severe price competition of the maturity stage.

Figure 5.4 also summarizes the different management goals over the life cycle stages, from an initial stage when management of innovation is most important (and managing quality, price, and so on, are less important), to management of rapid production, to management for maintaining quality and cost in the face of strong competition, to management for as little inventory as possible.

5.2.2 Production Strategy in the Product Life Cycle

Which production system should be chosen for manufacturing a product depends on the life cycle stage and the particular conditions of the product. Hayes and Wheelwright (1984) refer to a product and process matrix. Which production system should be chosen may be considered in terms of that, as illustrated in Figure 5.5.

A normal situation is the one in which there are low sales and unpredictable demand in the start-up stage. General purpose facilities and jobbing shop production are adopted for manufacture. In the rapid growth stage, with rapidly increasing demand and reducing variety of products, lot production with changing setups is adopted. In the maturity stage, line or continuous production with special purpose facilities satisfies the high-volume demands for a limited variety of products. Line production continues through the commodity stage, but if the product goes into decline, a return to job shop production may occur. In summary, as the stage of a product's life cycle progresses from left to right in the matrix of Figure 5.5, the production strategy mostly changes in a linear way from top to bottom.

A company may make the decision to force the pace of change. It may decide to anticipate the rapid growth stage and change to lot production during the start-up stage. This is the aggressive strategy in Figure 5.5. It may decide not to invest in lot production facilities until after the rapid

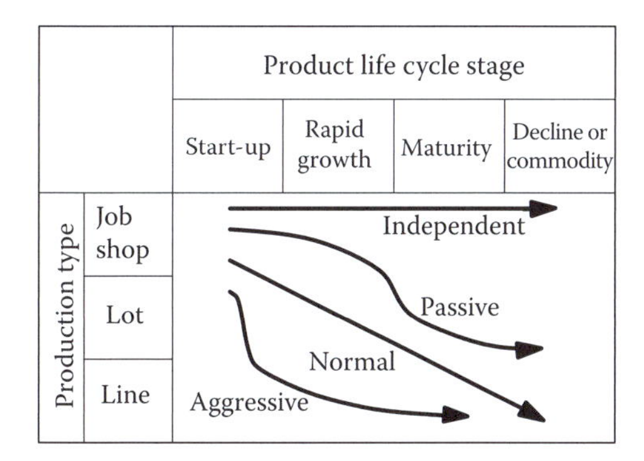

FIGURE 5.5
Product and process matrix and production strategies. (Developed from Hayes and Wheelwright, *Restoring Our Competitive Edge: Competing through Manufacturing*, New York: John Wiley, 1984, with additional material.)

growth stage has proved that the product is successful. This is the passive strategy. A new product may have a value for a very small market. Then investment in specialized facilities may never be justified. This is the independent case in Figure 5.5.

5.3 TECHNOLOGY TRANSFER AND MANAGEMENT OF TECHNOLOGY

The subject of this book is technology transfer. What technology is appropriate to be transferred depends on the product life cycle stage and production strategies that have just been described. It also depends on the manufacturing capability of the receiving side at different stages of its industrialization. This main section of this chapter expands on these, including the roles of management technology in appropriate technology transfer. It also considers strategic issues of technology transfer and management technology.

5.3.1 Appropriate Technology Transfer and the Role of Management

It is not always true that extending technology assistance to developing countries helps the national economy of the countries and makes economic development of the countries possible. Transferring the technologies developed in developed countries may not be effective for the economic development of developing countries.

Mass production and high-level incomes are the norm in developed countries. The capital-intensive technologies that have grown up in these countries are appropriate for them. If these capital-intensive technologies are transferred to a developing country, the transfer does not contribute to an increase in employment but leads to distorting the country's industrial structure and development. The technology transferred should be appropriate to the needs of the economy, society, and culture of the receiving side. Furthermore, in order to advance its economic development, the receiving side should introduce the technology itself, under its own control and by itself.

What is appropriate technology has been discussed all over the world. It is technology that contributes to the short-term development of the receiving side and also enables long-term and self-reliant development. It is also necessary to improve the ability of the receiving side to be inventive and develop technical innovation. The following paragraphs deal with what factors need to be taken into account in considering what is appropriate technology to be transferred for the development of the receiving side, under its own control and autonomously. There are four factors in all.

5.3.1.1 Importance of State of Development

The first factor is what already exists. This is not only what technology already exists, but also what is the potential of the receiving side to develop a technology and what systems are in place to support it. Even when the same technology is transferred to two different countries, the direct and indirect effects will differ. They will depend on the strength of the supporting technologies already existing in the country and on whether the receiving side is able to develop the technology properly. How well the introduction is organized is also important.

For these reasons, the condition of the receiving side must be taken into account as part of a technology transfer strategy. It is not enough just to enforce the transferring side's technology. The existing state of technology should be used as a foundation for new developments. The receiving side's environment should be developed to a state that is suitable to accept the transferring side's technology. In that way, opportunities for development can be expected to increase.

The differences in existing technology, potential for development, and systems in place between the transferring and receiving sides may be expressed all together as a difference in production structure. What is appropriate technology to be introduced depends on whether the receiving side's production structure is fully internally developed or is only partly developed with islands of foreign technology within it. In the first case, the production structure will already be able to produce a wide variety of cost-competitive products. A newly introduced technology, whether it is product, process, or management technology, will be understood, implemented, and utilized to develop the competitive power of the recipient side. In the second case, the effect of a technology transfer may be limited because of a poor relationship between the newly introduced technology and the existing technologies. Hence, more attention must be paid, when introducing technology, to the receiving side's production structure.

5.3.1.2 Importance of Human Resources

The second factor is the receiving side's human resources and education system. Even if the latest production facility or technology is introduced by a technology transfer, engineers and managers who are able to understand, utilize, and develop the facility or technology are essential to set it up and develop it. Thus, the capability of engineers or managers will affect what is appropriate technology. It will affect significantly the receiving side's ability to develop itself.

Therefore, in technology transfer, the education and training of engineers and managers must not be excluded. A critical issue in technology transfer is to organize an education and training system for engineers and managers so that they can understand, effectively use, and develop the introduced technology. Particularly, developing countries do not have enough engineers and managers. Of course, developing countries have engineers and managers who have received education in developed countries or who have a similar level of knowledge and skill to that. However, because they are not as many as needed, they are treated as an elite, and there is a tendency for them to distance themselves from the direct work of manufacturing production.

In Japan, the manufacturing education gap between manufacturing process workers and engineers or managers is not wide. Workers, engineers, and managers work together easily. The manufacturing and management technologies that support quality control activities with full worker, engineer, and manager participation are appropriate technologies that increase the international competitiveness of Japanese companies. Autonomous workers' groups, bottom-up decision making, and creating teamwork through job rotation are all able to be introduced in Japanese companies because employees have wide skills and knowledge. The Japanese management system supports this through such things as a lifelong employment system, seniority system, less laying off, and management with employee participation.

However, when there is a wide educational gap among employees, activities that assume the participation of all employees may not be effective. In this case, technology to support basic education and training of employees or assigning employees to particular jobs is considered as appropriate.

5.3.1.3 Importance of Market Competition

The third factor is the state of market competition. The size of the market, particularly the domestic market, influences the nature of the competition, and hence what is the corresponding appropriate technology for it.

As described in the section on product life cycle and process structure, the function of a product is most important at its start-up stage. Price and even quality are not so important, and neither is production volume nor delivery. Thus, appropriate technology at the start-up stage is not refined management technology aimed at cost competition, improved quality, and reduced stock, but management technology to produce the product with good enough quality and at an acceptable price. On the other hand, in the maturity stage, it is necessary to meet not only the functional product requirements but also the quality and price, volume, and delivery time requirements. At this stage, more refined management technology is considered as an appropriate technology.

An example of the importance of the state of the market comes from China. China has a population of more than 1.3 billion people. Its economy is developing in a bipolar manner with a big gap between urban and rural areas. Chinese household appliances have a huge domestic market. They are competitive in overseas markets, even though they may have low quality, because a favorable exchange rate enables them to be sold cheaply. When considering technology transfer to China, for Chinese household appliances companies, there is not a strong need for a refined management technology. At present, the Chinese household appliances industry is in the rapid growth stage, with drastic increase of consumption demand. Technology transfer is mainly about mass production technology. However, Chinese capability to develop part and product technology independently is still far behind that of other developed countries, like Japan.

For China's healthy industrial development, in addition to mass production technology, it is necessary for it to develop and innovate independently across all ranges of technologies, such as materials technology, part processing technology, and product design technology. It is not necessary at first to expect rapid development in these areas. Rather, improving the technical capabilities and accumulation of technology of individual companies, related companies, or a whole industry is much more important for China's longer-term healthy industrial development.

5.3.1.4 Importance of Strategic Factors

The fourth factor in what is an appropriate technology transfer is long-term strategy. An appropriate technology transfer is one that not only satisfies the direct and targeted short-term purpose of a company. In the longer term, it is able to satisfy the expectations of both sides in the transfer. A technology transfer that is appropriate only for one side or leads to profits for both sides, but only in the short term, and cannot expect development in the long run, is not desirable. A technology transfer that both sides see not only as a short-term target, but in the long run leads to good cooperative relations, is desirable. For these reasons, technology transfer is not only associated with profits for the transferring side, but also inevitably leads to the independent development of the receiving side.

5.3.2 Technology Strategy and Issues of Management Technology

It is common (for example F. Komoda, *Theory of International Technology Transfer* (in Japanese), Tokyo: Yuhikaku, 1987) to classify companies' technology strategies under the following six headings:

- Offensive strategy
- Defensive strategy
- Imitative strategy
- Dependent strategy
- Traditional strategy
- Opportunity strategy

This section explains more about these technology strategies and their implications for management technology. Companies that pursue these different strategies generally are involved in technology transfer at different stages of a product's life cycle, as summarized in Figure 5.6, developed from Figure 5.4. In each of the six parts of Figure 5.6 there is a horizontal timeline that for part of the life cycle is shown as solid and for part as dashed. The solid part denotes the time over which a company with a particular strategy is active in the life cycle, and the dashed part the time over which it is inactive. Technology transfer to or from the company occurs at the changeover time.

The next paragraphs consider each of the six strategies one by one, including implications for management resources.

5.3.2.1 Offensive Strategy

In an offensive strategy, companies carry out research and development activities in every type of technology by themselves, including fundamental research and product technology, as well as manufacturing technology research. As a result, they constantly lead technological innovation in their industry and make a monopolistic profit by controlling the market. Under this strategy, a large amount of resources in a company have to be allocated to technological development.

As a product enters the rapid growth or maturity stage of its life cycle, and the market changes from competition on function to competition on quality, price, and delivery (Figure 5.4), it becomes difficult to continue technical development aimed at quality improvement, cost reduction, and so on, because of lack of remaining resources. Thus, it is in the maturity stage, as shown in Figure 5.6(1), that companies with an offensive strategy, with decreasing competitive power, will transfer their technology to any company that may be able to produce with high quality and low price, aiming for profit from the acquisition of a licensing fee.

5.3.2.2 Defensive Strategy

A defensive strategy company also carries out various researches and development by itself, at first sight similarly to an offensive strategy company. However, it does not necessarily favor leading technical innovation. Rather, it targets its research and development more effectively by observing the failures of offensive strategy companies. In this strategy, the main focus is not original technology based on fundamental researches, but refinement of technology and design by experimental research.

In this way a defensive strategy company avoids development risks in the start-up stage of the product life cycle. It is to be expected that its cost of research and development is reduced from that in an offensive strategy, even though the cost of development may be high in order to gain a market share. The financial resources saved by reducing risk can be used offensively for management technologies during the maturity stage, to improve quality and cost reductions and to maintain competitive power for longer than can an offensive company. Because of its later entry into the market, if a defensive strategy company were to leave the market early, in the same way as an offensive company, its total profit would be reduced. Hence, as shown in Figure 5.6(2), a defensive strategy company does not transfer its technology until the commodity or declining market stage.

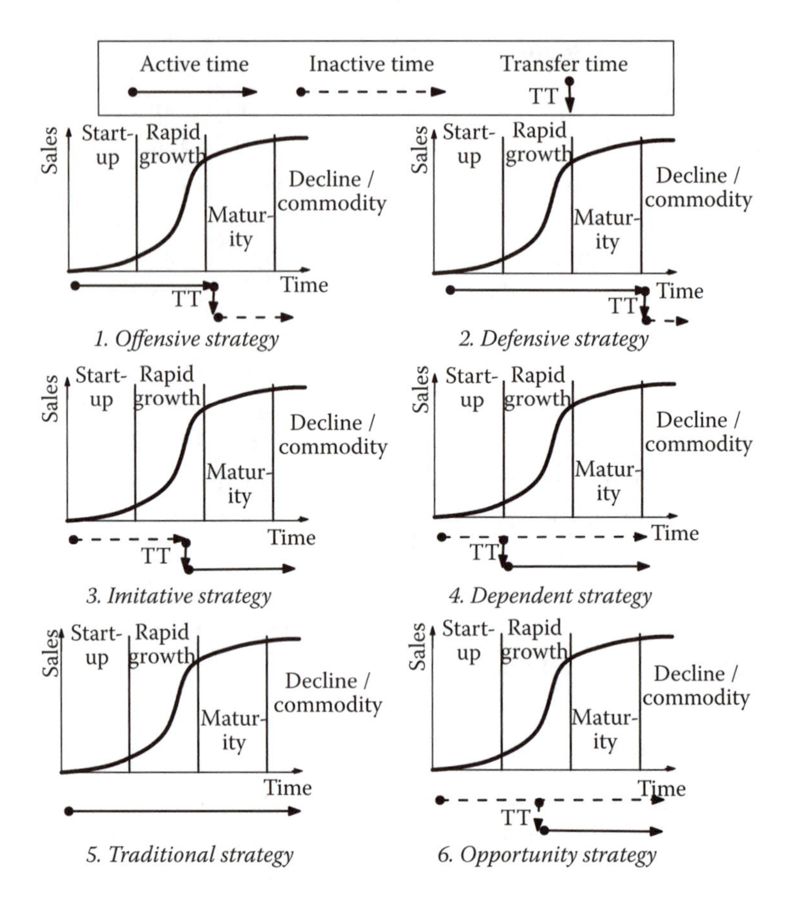

FIGURE 5.6
Technology strategies within the product life cycle.

5.3.2.3 *Imitative Strategy*

Companies that pursue an imitative strategy mainly take up technologies developed and established by others. Research and development under this strategy is focused on incidental functions of the product, such as quality improvement and cost reduction by improvement of the production process, technical service, or training. Once a product's maturity stage has been reached, and market competition is based on product quality and price, companies with an imitative strategy become competitive by supplying high-quality and low-price products, as well as after-sale service, all based on having an effective production process. As shown in Figure 5.6(3), companies with an imitative strategy tend to be receivers of technology transfer toward the end of the rapid growth period of the life cycle.

Companies with an imitative strategy aim to introduce new technology and effective production methods to produce high-quality products at a cost much less than that of the existing technology or that the market asks for. Development of product technology is not of interest, but production organization is called for that can survive the severe product quality and price competition. It can be imagined that the total profit obtained from market entry by an imitative strategy company is limited because of severe market competition and late entry into the market compared to offensive or defensive strategy companies.

5.3.2.4 Dependent Strategy

In a dependent strategy not only research and development but also imitation and technical improvement are abandoned. A dependent strategy company relies on being strongly linked to a company with technological predominance. As shown in Figure 5.6(4), it receives technology from the predominant company at or soon after the start-up stage of the life cycle. Thereafter, it supplies parts and services to the predominant company.

A company that has a dependent strategy, which never improves its production, or the parts and services that it has been asked to supply, depends for its survival on its customer company. It cannot be said to be independently sound. For independence, some effort at technology development for improving the specification of the parts and services allotted and efforts for improving productivity are indispensable.

5.3.2.5 Traditional Strategy

A traditional strategy seeks development based on craftsman skills, for example in precapitalistic social systems or in small, divided, regional markets. It is difficult to apply technology transfer as part of this strategy, as almost all ability lies in the traditional skills of the workers. Thus, in Figure 5.6(5), no technology transfer is shown.

In this situation, in which everything, including product development and manufacturing technology, is held by the craftsman as skill and know-how, there is little explicit knowledge. The management task, in supporting continuity and technology development, is to create an environment that supports individuals in handing down their tacit knowledge to the next generation.

5.3.2.6 Opportunity Strategy

Opportunity strategy needs neither research nor development nor improvement. It is a strategy that looks for new opportunities particularly in rapidly changing (and growing) markets. Thus, as shown in Figure 5.6(6), it tends to start from the middle of the rapid growth stage of the life cycle. Because this strategy tries to find anything to connect with market opportunity, without research and development, marketing management is important in this strategy.

5.3.3 Strategic Technology Transfer and Sustainable Development

The financial resources of companies are limited. It is by appropriate technology transfers, suited to the different strategies of the companies or organizations involved, that these finite resources are used most effectively to increase wealth. Appropriate technology transfers give the best chance of sustainable development. Relations between the technologies and the strategies have been described in the previous sections.

The companies or organizations whose resources should be used most effectively are those transferring as well as receiving technology. At the present time, various new technologies are being developed against a background of limited global and economic resources and requirements to consider environmental effects. In the future, technology transfer between companies and company strategies will continue with the aim of sustainable development within limited resources. The transfers and strategies are likely to broaden, to include collaborative relationships with other partner companies and organizations. They are likely to broaden further to contribute to the development of the surrounding societies. It is with such a broad view that the most effective and sustainable transfers will occur.

DISCUSSION QUESTIONS

1. What are the three elements of demand? For each, describe the factors that affect their importance and give examples.
2. Describe the effects on performance of a company of changing production strategies at different stages of the product life cycle.
3. For each technology strategy of Section 5.3.2, give examples of companies that utilize that technology strategy and state the reasons.

6

Overseas Expansion and Technology Transfer

After the collapse of the bubble economy in the early 1990s, Japan entered a long continuing recession. However, through various efforts by companies, domestic business recovered. After a lowest point in 2002, it eventually surpassed the levels seen in the miraculous expansion (the Izanagi Keiki) of November 1965–July 1970. One effort was the shift of manufacturing by Japanese companies to overseas, especially to South East Asia and China. Companies took this risk in order to survive. They invested in building an overseas manufacturing base and in advanced technology transfer in order to gain reduced costs through cheap labor.

Manufacturing overseas not only gave a direct reduction in production costs, but it also supported improved company profits through market expansion opportunities. But once domestic business started to recover, it became an aim to stimulate stagnant domestic industry. The domestic recovery of manufacturing technology through domestic location of manufacturing industry and prevention of advanced technology transfer overseas began to be discussed as part of the global strategy of companies.

Today Japan continues to face an aging society and population decline. This is not favorable to domestic market expansion, as there is seen to be a limit to domestic demand. Domestic demand can be stimulated by new industries developing new high-value-added goods to create new markets. But, it is only a matter of time before expansion overseas again becomes important.

In the case of overseas expansion, unlike domestic expansion, differences between the overseas and domestic environment need to be considered, for example exchange rates, the legal system (tax system, regulation, etc.), infrastructure, labor costs, and local talent. In the global strategy of companies, the most important thing is how to move domestic technology smoothly overseas and to carry out production activities as effectively as domestically.

In this chapter some topics are covered that the reader should appreciate.

6.1 SPECIAL FEATURES OF TECHNOLOGY TRANSFER OVERSEAS

Technology transfer overseas refers to transfer from Japanese domestic companies to Japanese subsidiary companies or local companies overseas, for example the transfer of manufacturing technology from Japan to countries in South East Asia and China. Technology transfer has been defined by T. Ando et al. (*Technology Development and Technology Transfer in China* (in Japanese), Kyoto: Minerva Shobo, 2005) as the transfer of the technology of one economic unit to another economic unit, for the same purpose and where the production activity is carried out. This is generally known as technology transfer between subjects. Technology transfer can be between various subjects, for example between companies, generations, or a university and a company. Overseas technology transfer is transfer from a domestic subject to an overseas subject. Differences in the cultural backgrounds of the transferring country (Japan) and receiving countries (for example, developing countries such as in South East Asia, China, and so on) play a large role. These differences include differences of language, country conditions, ability to absorb technology, and the level of technology. Without these differences, there is no difference between overseas technology transfer and domestic technology transfer.

In technology transfer, information related to manufacturing is broken down into different parts. These are transferred to personnel on the receiving side by means such as documents, pilot equipment, and direct communication by experts. They are then combined again to re-create the whole technology. In the case of technology transfer overseas, it is difficult to transfer to a receiving side that has a different cultural background, country conditions, education level, and technical level. Thus, careful preparation is necessary for overseas technology transfer.

According to Y. Okamoto (*Japanese Company Technology Transfer Assimilation in Asian Countries* (in Japanese), Tokyo: Nihon Keizai Hyoronsha, 1998), problems of overseas technology transfer can be considered in terms of two factors: technology gap and cultural difference. These can further be divided into factors that can and cannot be controlled by the transferring side. Although controllable factors can be overcome by the efforts of the transferring side, uncontrollable factors, originating from the societal and economic environment of the receiving side, cannot be. Cultural background, country conditions, and education level are all uncontrollable factors. The transferring side should fully understand this in order to make progress in technology transfer.

The success or failure of overseas technology transfer is related to the level of understanding and the ability to absorb technology on the receiving side. If the transferring side understands well the conditions of the receiving side, and if the receiving side has a good ability to absorb technology, the transferring side's system can be transferred simply. If this is not the case, it becomes necessary to convert the transferring side's (Japanese) system to a system that is suited to the receiving side, taking its environment into account. It is also necessary to create a logical technology transfer structure that is suited to the receiving side.

6.2 HISTORICAL BACKGROUND TO OVERSEAS TECHNOLOGY TRANSFER

After World War II, Japan brought in overseas technology. Economic growth came through the export of products based on that technology. Revival and increased GDP were achieved. There is no doubt that this economic growth was helped by the fixed exchange rate ($1 = 360 yen) that benefited exports. When a floating rate of exchange was introduced in January 1976 and the yen started to appreciate, the auto industry protected itself by developing overseas footholds while other companies concentrated on cutting costs. In 1985, there was further appreciation of the yen as a result of the Plaza Agreement between France, West Germany, Japan, the United States, and the UK. As a result of this and other factors, Japan's manufacturing industry slowly lost its international competitive power. Other factors were the increased wages and quality of life brought about by economic growth, and the start of the decrease in young workers as a result of a low birthrate and increased life expectancy. All these led to increased manufacturing costs and a loss of competitiveness with overseas products. Products such as textiles and household appliances began to be produced in South East Asia as a result of technology transfer. Why many products became imported to Japan is easy to understand.

These are the reasons why Japanese manufacturing industry, which grew after the war because of a strong manufacturing culture, had to move its foothold overseas to cheap labor regions like South East Asia and China. However, a consequence of moving production overseas was the decline of domestic manufacturing industry. To counter this, domestic manufacturing industry shifted to high-value-added goods and carried out operating reforms, including restructuring. These domestic reforms did not always go smoothly and were affected by the long-term economic slump. From the 1990s measures and deregulation were carried out by government. Companies were further encouraged to develop high-value-added goods and create new industries. University-led ventures were founded to harness university intellect.

Technology transfer to developing regions in South East Asia and to China, including direct investments, progressed steadily during the economic recession. Access to cheap labor and reduction in manufacturing costs were the primary objectives. Technology transfer started with Korea and Taiwan, and then proceeded to South East Asian countries like Indonesia and Thailand. Today technology transfer to China is well known. Furthermore, technology transfer to Vietnam has progressed, and in the future there might be a big move to Laos and Cambodia and perhaps even India. Today the expanded background of the country that is transferred to means that production is not moved overseas only to obtain low labor costs, but for market expansion as well. In particular, increasing numbers of companies are shifting to China. This is not only for cheap labor, but also because of possible expansion to the attractively huge market of 1.3 billion people.

6.3 OVERSEAS EXPANSION AND CONDITIONS OF TECHNOLOGY TRANSFER

For overseas expansion by Japanese companies, there are some processes that need to be gone through, from setting up advance planning to the start of production and sales. Detailed steps of overseas expansion have been summarized, especially for the case of developing countries, in S.

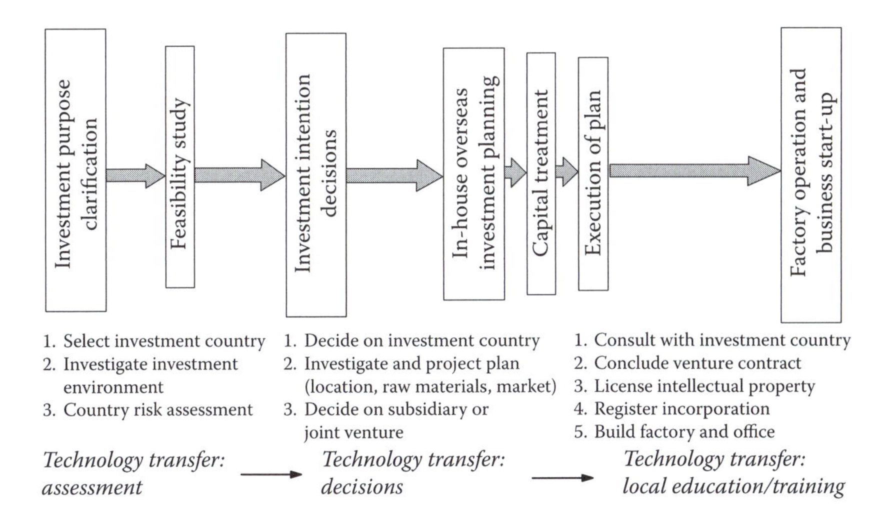

FIGURE 6.1
Manufacturing industry advance overseas planning process. (Based on material in S. Tsuchya, *Company Expansion to Asia and Overseas Job Transfer: Planning and Execution* (in Japanese), Tokyo: Nikkan Kogyo Shimbun, 1995.)

Tsuchya (*Company Expansion to Asia and Overseas Job Transfer: Planning and Execution* (in Japanese), Tokyo: Nikkan Kogyo Shimbun, 1995). Figure 6.1 gives a step-by-step view. A purpose and feasibility assessment is followed by a decision to invest and then by the transfer and training of local employees. In more detail steps are: clarifying reasons for transfer → carrying out feasibility study → decision on investment intention → in-house facilities and financial project planning → raising the capital → executing the plan → factory operation and business start-up.

In the following paragraphs, overseas expansion and technology transfer are considered as companies see it, including results from a survey of companies with experience in overseas expansion, undertaken in 2006 by Hiroshima University. The study was funded by the Japanese Ministry of Economy, Trade and Industry.

6.3.1 Strategy in Technology Transfer

The globalization of Japan's manufacturing industry started as a result of the collapse of the bubble economy. Mainly manufacturing industry with large labor requirements moved to areas with cheap labor, like South East Asia and China. However, the strategy of transferring technology overseas should be considered more widely than from the point of view of labor costs. The list of strategic reasons below starts with labor costs but continues with two other items:

- Finding cheap labor costs. Transfer of technology to reduce manufacturing costs is the goal. Textile, information and telecommunications, electrical goods, and precision instrument industries provide some typical examples.
- Responding to the request of a parent company to move production and transfer technology after the parent company has itself built production facilities and transferred technology overseas. To make parts locally is the goal. This is a common occurrence, particularly with large automakers and their suppliers.
- Transferring technologies and moving manufacturing to seek new markets. Market expansion is the goal. Production is sought in an optimal location with good sales opportunities.

In all cases the purpose is to increase sales. In the case of the first bullet, it is by exporting locally made goods overseas. For the second bullet, it is by supplying parts to the parent company's overseas site. In the third case, it is generally to increase market share, including by exporting back to Japan.

Table 6.1 shows personnel costs in the Asian region, gathered by the Japan External Trade Organization in 2006. Compared to personnel costs in Japan, those in the Asian region are lower. The reduced personnel costs are closely related to the much lower manufacturing costs in the region.

However, the receiving side in technology transfer must gain from the transfer as well. Rather than just taking lower personnel costs and decreased manufacturing costs from the receiving side, the transferring side that aims to raise its global competitiveness should contribute to the development of the developing country as well, to improve the country's technical capabilities, increase its employment, and its economic development.

It is only natural that the strategic reasons for technology transfer differ between processing industries, for example raw materials industries such as textiles, textile processing, semiconductor and parts manufacturing industries, and assembly industries, for example car and electrical machinery industries.

TABLE 6.1

Labor Costs in the Asian Region, $/Month Converted from Local Currency at the Rates Shown

Nation/City	Japan/ Yokohama	China/ Beijing	China/ Shenzhen	China/ Shanghai	Korea/Seoul	Thailand/ Bangkok	Malaysia/ Kuala Lumpur	India/New Delhi	Vietnam/ Hanoi
Currency (for $1)	117.72 (yen)		7.8715 (yuan)		941.70 (won)	36.525 (baht)	3.6403 (ringgit)	45.34 (Indian rupee)	16.093 (Dong)
Worker	3090	83 ~ 264	123 ~ 509	272 ~ 362	1573 ~ 1691	164	221	165 ~ 326	87 ~ 198
Engineer (middle)	3637 ~ 4200	178 ~ 330	194 ~ 794	441 ~ 641	2175 ~ 2279	383	820	394 ~ 799	243 ~ 482
Manager (middle)	4640	356 ~ 1003	528 ~ 1060	663	3315 ~ 3489	684	1638	696 ~ 1684	597 ~ 859

Source: Data from 16th Survey of Investment-Related Cost Comparison in Major Cities and Regions in Asia, Japan External Trade Organization, 2006, in Japanese.

6.3.2 Statistics of Overseas Expansion

Many studies have been made in North America, Europe, Asia, and elsewhere of how much Japanese companies have expanded overseas. The Japanese Ministry of Economy, Trade and Industry has carried out Overseas Business Activity Basics Investigations since 1971. It knows well the actual condition of globalization of Japanese companies. Data have been released 36 times so far. In the *26th Survey of Overseas Business Activities* (March 20, 1998), the number of Japanese companies expanding overseas is recorded as reaching a peak of 424 companies in 1988, before reducing to a low point in 1992 with the collapse of the bubble economy, and then afterwards starting to increase again. On the other hand, the number of companies withdrawing from overseas was 107 in 1993 and 108 in 1994, before decreasing sharply to 43 in 1995. Figure 6.2 shows statistics for manufacturing industry for the 11 years from 1995 to 2005, based on the 35th and 36th Overseas Business Activity Basics Investigations. After the peak in 1995, the number of company expansions reduced to a lowest value in 1998, before increasing again to reach another peak in 2002, and decreasing again after that. Over the same period, the number of overseas withdrawals increased from a low in 1995 to reach a peak in 2002, before slowly decreasing again.

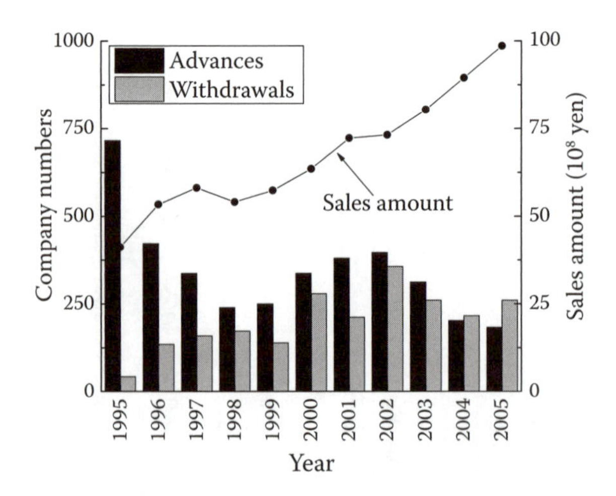

FIGURE 6.2
Changing company numbers overseas (advances and withdrawals) and total sales value for the period 1995 to 2005. (Data from *35th and 36th Surveys of Overseas Business Activity*, Ministry of Economy, Trade and Industry, Japan, 2006 and 2007.)

It is understood that the number of cases of Japanese companies' expansion overseas is influenced by external factors (collapse of the bubble economy, change in the value of the Thai baht as a result of a floating exchange rate, business trends, etc.). Although the number of withdrawal cases also depends on economic changes, it is more closely related to company management internal factors (for example, miscalculation of sales). In Figure 6.2 changes in expansions and withdrawals are seen every year, but because there are more companies expanding than withdrawing every year, it can be said that overall the number of Japanese companies expanding overseas increased year on year.

An exception was in 2005, when the number of withdrawals was higher than the number of expansions. According to investigations, the reasons for this were reorganizations of companies due to financial reviews, leading to company integrations and strategic withdrawals. Although the number of cases of expansion and withdrawal changed, the amount of sales increased every year, demonstrating the importance of manufacturing overseas.

How overseas expansion has varied from region to region (mainly North America, Asia, and Europe) is shown in Figures 6.3 and 6.4 and Table 6.2. Figure 6.3 shows the Japanese overseas production ratio from 1995 to 2006. Overseas production ratio is the ratio of Japanese overseas production to the sum of home and overseas production. Figure 6.3 shows it to have expanded from 8 to 18% from 1995 to 2006 and that, within that increase, the percent

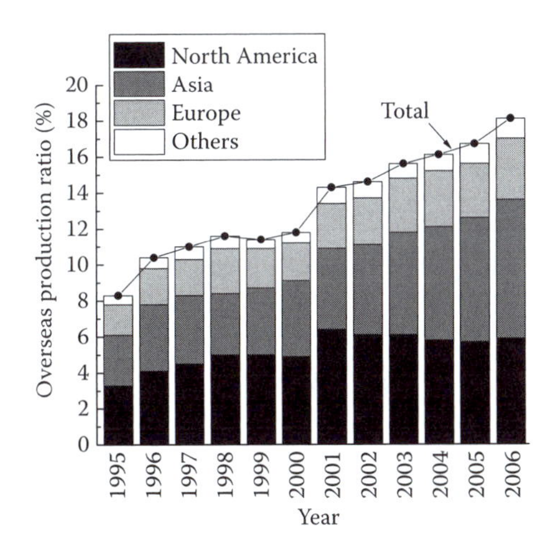

FIGURE 6.3
Overseas production ratio by year and region. (Data from *35th Survey of Overseas Business Activity*, Ministry of Economy, Trade and Industry, Japan, 2006.)

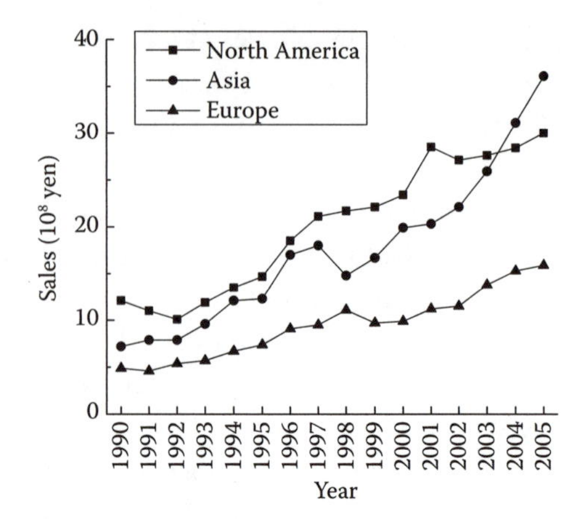

FIGURE 6.4
Sales by overseas subsidiary companies by year and region. (Data from *36th Survey of Overseas Business Activity*, Ministry of Economy, Trade and Industry, Japan, 2007.)

accounted for by the Asian region rose from 28 to 42%. The increasing importance of the Asian region is also seen in Figure 6.4. Sales revenue from the Asian region overtook that from North America in 2004.

Table 6.2 shows the number of overseas subsidiary companies by region. It also shows a steady increase in the Asian region, and that that increase

TABLE 6.2
Share of Overseas Subsidiary Companies by Region, 2001 to 2005

Region		Subsidiary Company Numbers (2005)	Share by Region (%)				
			2001	2002	2003	2004	2005
North America		2825	20.8	20.0	19.0	18.9	17.8
Asia		9174	50.9	52.6	54.0	56.4	57.9
Made up of	China	4051	17.8	19.6	21.4	23.8	25.6
	ASEAN4[a]	2715	17.8	17.8	17.6	17.4	17.1
	NIEs3[b]	2044	12.9	12.9	12.7	13.0	12.9
Europe		2384	17.2	16.9	16.8	15.8	15.0
Others		1467	11.1	10.5	10.2	9.5	9.3
All regions		15,850	100	100	100	100	100
BRIC countries[c]		3502	15.0	16.5	18.5	20.4	22.1

Source: Data from *36th Survey of Overseas Business Activity*, Ministry of Economy, Trade and Industry, Japan, 2007; the last two rows are summary data.

[a] Indonesia, Malaysia, Philippines, Thailand.

[b] Republic of Korea, Singapore, Taiwan.

[c] Brazil, Russia, India, China.

is mainly accounted for by China. It also shows a steady increase year on year of subsidiary companies in Brazil, Russia, India, and China (the BRIC countries). These regions are expected to increase in importance in the future.

6.3.3 The Content of Technology Transfer

What is involved in overseas technology transfer depends on the type of manufacturing industry and the area to which it is transferred. The shifting of manufacturing industry with large workforce needs to areas with low labor costs, such as South East Asia and China, has already been mentioned. Although it is still the case that high-value-added technology production is mainly carried out domestically, it seems that this may currently be changing.

Technology transferred overseas can be divided into three types. One is product technology (technology related to a product such as design technology, compounding technology for polymers, etc.). Another is manufacturing technology (know-how that makes manufacturing possible and operating technology like machine tool technology). The third is management technology (technology for manufacturing management). According to the companies that were surveyed to obtain the information in Section 6.3.1, product technology was the subject of transfer in 71% of cases, manufacturing technology in 92%, and management technology in 86%. Although product technology is the least transferred, all these percent values are high. It seems that a subsidiary's operation cannot be carried out without such high values.

Figure 6.5 shows the extent to which the companies transferred either their own special technologies or only general technology, i.e., widely known technology. In this figure, each of the five headings, such as "special technology" and "general technology," is independent of the others. Taking the responses on production technology (gray bars) as an example to illustrate how the figure should be interpreted, 4% of the companies did not reply about production technology. Of the 96% that did reply, 78% indicated they transferred their special technology that they had also used domestically, and 32% that they had transferred general technology. The total exceeds 100% because some companies had transferred both types of technology.

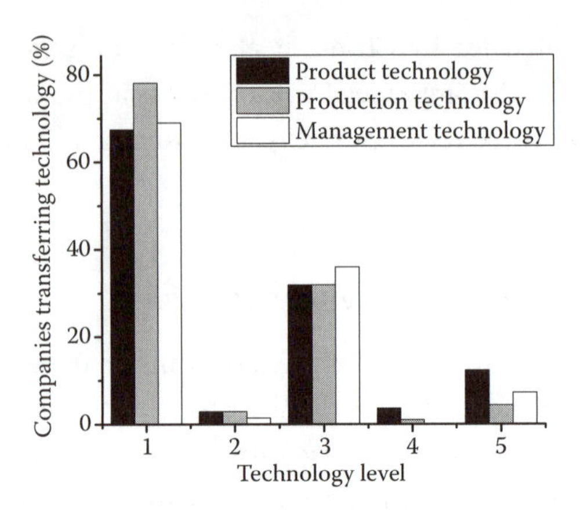

FIGURE 6.5
Companies transferring technology overseas, by technology type (product, production, and management) and level (1 = self-developed special technology, previously used domestically; 2 = as 1 but not used domestically, 3 = general technology, 4 = others, 5 = no response).

6.3.4 Important Considerations in Overseas Technology Transfer

Overseas technology transfer is different from technology transfer within domestic companies, as transfer is to a place with a different cultural background, customs, and language. Language in particular is a big obstacle. A large effort is needed to convert documents from their original form in the transferring side's (Japanese) language to the local language. Of course, this statement only applies to the transfer of explicit knowledge. Documents do not exist for implicit or tacit knowledge. It is difficult enough to transfer this domestically, without language problems. It is even more difficult to transfer it overseas. There is a major need to find better ways to support transfer of tacit knowledge across languages. After this, other issues also need to be considered.

What these other issues are, according to the surveyed companies, is summarized in Figure 6.6. The uncontrollable factors, already mentioned, such as the temperament of the workers (how honest, meticulous, obedient, stubborn, etc., they are), their education level, and lifestyle, are among the most mentioned issues.

Companies also commented that intellectual property protection is an issue. Figure 6.6 has this as the third most important problem. The fact that South East Asian countries and China do not much enforce

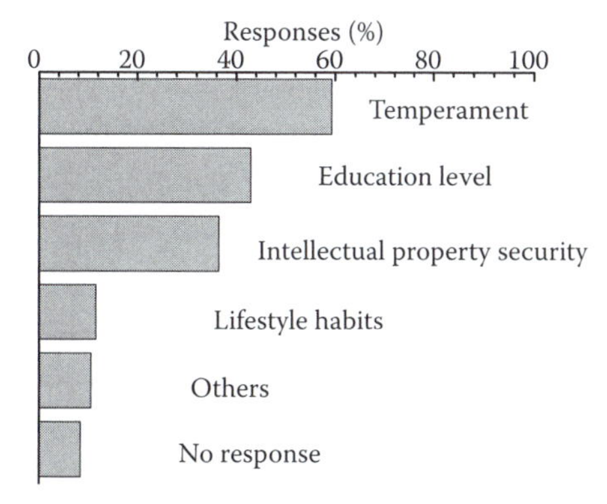

FIGURE 6.6
Technology transfer problems with the receiving side. (From answers by 189 companies.)

intellectual property law is certainly a factor that companies strongly think about.

Among other issues, a fast turnover of trained workers can also be a problem with established technologies, where other companies may offer competing job opportunities. Changing companies in order to gain better reward is a natural thing overseas. To counter this, companies should probably create special positions for the persons (the key persons) who are essential to the companies' success.

Although it might be possible for the transferring side to improve its capabilities as far as the controllable factors of transfer are concerned, the uncontrollable factors on the receiving side remain of most importance, particularly the learning abilities and technical level of the workers. As is implied in the previous paragraphs, how well the transferring side can train the workers on the receiving side determines the success of technology transfer.

As part of the transfer, work standard documents are normally prepared in the local language (or at least in English), and key persons on the receiving side undergo 3 to 6 months of training to master the technology. As written before, the main issue is language. Although generally English is used, it goes without saying that the local language is the best. The key persons, trained in the transfer of the technology, will themselves train the main workforce on the receiving side. The most difficult thing in technology transfer is to pass on know-how that cannot be documented in work standards (so-called tacit knowledge).

It is difficult to learn know-how in off-the-job training. Normally, therefore, personnel from the transferring side are sent to the local area to give the local workers on-the-job training. However, care must be taken, as possible language problems at that stage may cause difficulties. Training of workers on the receiving side is the most important part of technology transfer.

6.3.5 Procedures of Technology Transfer

Technology transfer is carried out by means of workers, facilities, and information. Generally, technical information is the most important. Information on product technology is found in various specifications (for example product design specifications, materials/parts specifications, and software specifications), design drawings, circuit diagrams, and so on. Facilities' inventory, facilities' operating methods, process organization charts, manuals on production management, and the transferring side's personnel structures are some of the examples of information on manufacturing technologies to be passed on to the workers on the receiving side. Although it is best for documents to be in the local language, English is the minimal requirement.

Technology transfer's dependence on the education level and technology absorption abilities of the receiving side has already been written about. In general, a transfer schedule is drawn up. The progress of transfer is based on that. Although it is best for training to be in the language of the receiving side, namely, the local language, in practice, first training is carried out with key persons who can converse in English. Then these key persons train the local people. The training of the key persons by means of the English language normally takes place in Japan. They then transfer the manufacturing skills to the local people in the local language. The biggest problem here is communication.

6.4 FUTURE TRENDS IN OVERSEAS TECHNOLOGY TRANSFER

Japanese industry is now clearly profiting from its overseas investments. As has been written before, in the 1990s overseas transfers started to South East Asia and China. The purpose was to lower manufacturing costs through cheap labor and thereby to increase competitiveness. Now a more global strategic view prevails. The number of manufacturing companies transferring operations overseas is steadily increasing. The amount of investment in North America, Europe, and Asia has equalized. The importance of Asia has become greater, and there is a continuing trend of transfers there. Overall, overseas sales and profit continue to increase, as shown in Figures 6.2 and 6.4.

With the declining birthrate in Japan, and the growing proportion of old people and eventual population decline, expanding overseas markets is the only way to overcome the upper limit to domestic demand. This view is seen in companies' responses to questions on why they expanded overseas. Companies were asked for their reasons for expanding at the time that expansion occurred and their reasons now. The responses in Table 6.3 are mainly under the headings cost reduction, customer requests, own judgment, and market exploitation. The decline in cost reduction and increase in new market exploitation as reasons for expansion overseas are clear.

At the same time, there is talk in Japan of a strong revival of domestic manufacturing in the future. Because of possible future problems overseas that might lead to loss of or withdrawal from overseas operations, and to prevent loss of core skills at home, new domestic factories should be built. Part of this argument is also that advanced technology should not be transferred overseas. Companies must pay attention to these views.

A view that takes all concerns into consideration is that production should be carried out wherever is the most appropriate place. As well as taking account of a strong yen that would tend to force production overseas, a longer-term view of domestic needs is also necessary. The results of the companies' surveys asking opinions about future trends in both domestic and overseas production are in Figures 6.7 and 6.8. Each gives percent values to the number of companies giving particular reasons for where they would manufacture in the future. Manufacture of high-value-added products and increasing production capacity for the domestic market are

TABLE 6.3

Objectives of Overseas Transfer at the Time of Expansion and Now, from Company Responses

Objective	At Time of Expansion (%)	Now (%)
Cost reduction	40.8	31.7
Customer requests	19.1	8.6
Independent judgment	18.9	15.4
New market exploitation	14.0	30.5
Others	7.3	13.8

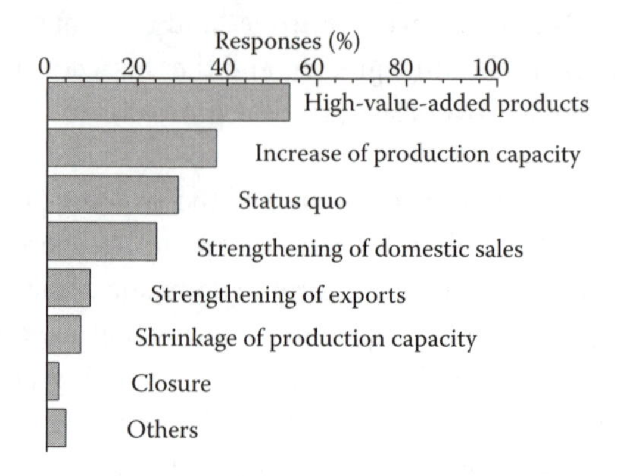

FIGURE 6.7
The company views of future trends in domestic production.

the two main reasons given for domestic manufacture. Increasing general purpose production capacity and strengthening local sales are the two main reasons for overseas manufacture. These views are expected to spread among Japanese companies.

As far as domestic manufacture is concerned, increase of production capacity is seen as necessary for strengthening domestic sales (the fourth most common reason in Figure 6.7). But for overseas manufacture a trend to high-value-added products comes next to strengthening local sales (Figure 6.8). Japanese companies believe that foreign companies will compete with them for high-value-added product markets, so they must respond in the overseas location. This has implications for the development of the labor force in the overseas locations.

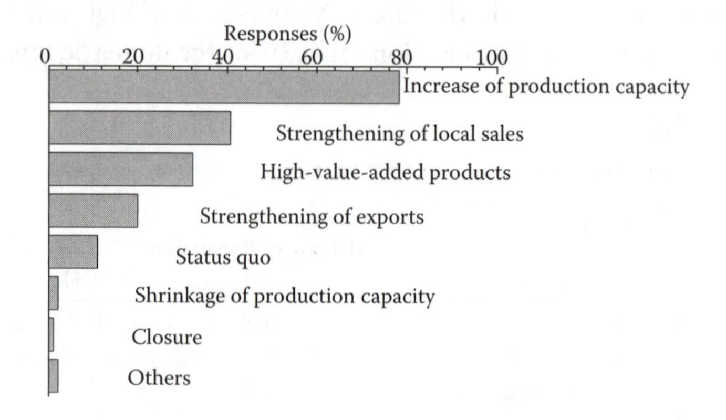

FIGURE 6.8
The company views of future trends in overseas production.

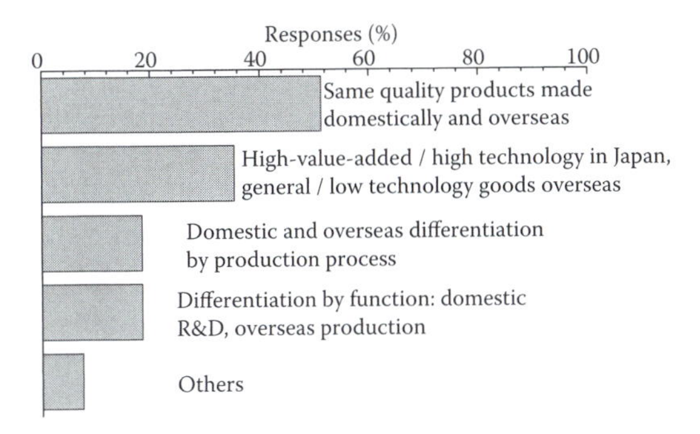

FIGURE 6.9
The company views on future division of labor.

The company views on how work will be divided between domestic and overseas sites in the future is summarized in Figure 6.9. Fifty-one percent think production will be duplicated, with the same products and quality, at home and overseas. Thirty-five percent think high-value-added goods will remain manufactured at home, while overseas production will concentrate on general (closer to commodity) production. Eighteen percent think there will be differentiation by production processes at home and overseas. Eighteen percent also think that research and development will remain at home, with production overseas.

It is clear from the survey answers that the places where companies have expanded overseas have become stronger, the role of exporting back to Japan has diminished, and there is a tendency for domestic companies and their overseas subsidiaries to manufacture the same products. Nonetheless, there is still a division that high-value-added products and high-technology products should be manufactured domestically and general products overseas.

Opinions that in the future some parts of a product will be produced domestically and others overseas are rather few in percentage terms (18%). It is not a matter of policy, but of sourcing from wherever is best with respect to manufacturing cost and quality.

As far as companies that manufacture the same products at home and overseas are concerned, the internal competition is felt to be a strength that will help competing with new external markets such as China.

Also, it is recognized that to produce the same products overseas as in Japan, it is necessary to transfer the Japanese technology overseas. The costs of lost sales opportunities by not transferring the technology exceed the risks.

However, there are some cases in which transferring advanced technology cannot work, and then the traditional pattern of manufacturing high-value-added and high-technology goods domestically and general products overseas remains in place. These cases are where the technical level of workers in a country is not high enough to work with the advanced technology.

DISCUSSION QUESTIONS

1. It is said that overseas production leads to a stagnation of domestic enterprises. It is required to balance overseas and domestic production. Describe your ideas, from a business manager's point of view, on what business maneuvers or technology strategy you should take.
2. In order for smooth technology transfer from one country to another, it is important to consider uncontrollable factors, such as technological and cultural differences between the transferring and receiving sides that the transferring side cannot manage. Describe your thoughts about concrete countermeasures.
3. Describe your opinion on how methods of technology transfer to other countries will change in the future, recognizing that technology transfer differs by region (United States, Europe, and Asia) and the technology field.

7

Technology Transfer and Legal Affairs

Before every technology transfer the transferring and receiving sides must decide what is to be transferred and how to do it. The outcome is an agreement. Without an agreement nothing will go forward. Small agreements between private individuals can be gentlemen's agreements. Gentlemen's agreements rely on honor and trust between people for them to be carried out. Large agreements that involve many people and resources must be more formal. It is too risky for them to rely only on honor and trust. A legal agreement is a formal agreement between the parties that is drawn up within a framework of laws within which both sides agree to be governed.

A well-thought-out agreement will cover all aspects of the technology transfer. All aspects are not only all the necessary activities and resources, but also how to proceed if problems occur. The more clear an agreement is, and the more its content can be referred to by people to help them carry out the transfer, the more likely is the transfer to proceed smoothly, to reach a successful conclusion. Specialists in legal affairs are skillful in how to draw up agreements. They know from experience what items should be included and where problems may occur.

Section 7.1 briefly describes what is the function of legal affairs in technology transfer. What is ethically to be expected from legal affairs specialists is included here. Section 7.2 covers what are the range of activities that should be agreed. Obviously this includes what is to be transferred. But licensing, staffing, construction, and procurement are all among details of how to proceed that should be decided on as part of an agreement. Section 7.2 covers these things by means of an example outline agreement. There are many details in legal agreements that people who are not legal affairs specialists often do not properly understand or think are not important. As a result, they may accept aspects of an agreement that later they regret. Such things are the subject of Section 7.3.

7.1 FUNCTION OF LEGAL AFFAIRS IN TECHNOLOGY TRANSFER

The term *constitutional state* may be familiar. A simple definition is that a constitutional state is "a state which takes as its principle the use of national power based on law (this is the principle of constitutional government)." Those who belong to the constitutional state (individuals, corporations, etc.) obey its laws and ordinances (i.e., the rules that are established by the executive branch), and in return they are eligible for protection.

Most countries in the world today are constitutional states, although there may be a few differences between their laws and ordinances. In these cases, when carrying out technology transfer to countries overseas, besides the laws and ordinances of the home country, the laws and ordinances of the transfer partner country relating to technology are usually of concern too.

In the case of agreements on technology transfer with foreign countries that have different cultures, languages, customs, and laws, the agreement that is finally reached is normally summarized in a common language (usually English) after passing through the negotiation stage in various forms. The specialists in legal affairs are the ones in charge of the series of negotiations. The legal affairs specialists are lawyers, including public relations lawyers who are well aware of overseas relations. In cases of companies, there are instances where in-house lawyers or persons in charge of legal affairs are employed who do not hold a lawyer qualification but are well experienced in foreign affairs.

How should specialists in such legal affairs handle technology transfer cases and from which point of view? First, the legal agreement shall be established between both parties. Within both countries, there will be their own domestic laws and treaties concerning international relations. The job of the legal affairs specialists is to build a legal agreement relationship within this framework. They should avoid illegal activities, and they should not look for legal loopholes or even suggest them. Also, they should not be required or expected to look for such things.

Next, in a legal arrangement, it goes without saying that it is only natural for the legal affairs specialists to produce efforts that conclude agreements that are beneficial to their employers. A conclusion is brought about as a result of an "intellectual battle" between the specialists in legal affairs on each side.

However, the conclusion of the agreement is not the end. At the time that the conclusion is reached, both parties start implementation. This is not necessarily trouble-free. Sometimes there may be differing opinions on the implementation of an agreement. Different interpretations of the clauses of an agreement could occur. Sorting out these problems at the same time as the process of implementation is continuing on both sides is not that easy. Depending on the case, it is not just a matter of settling the problem with the party concerned, but carrying it out according to the process for dispute settlement as determined earlier in the agreement. This is where legal affairs specialists appear again in this agreement implementation process.

In companies, legal affairs, unlike line functions such as manufacturing and product sales, are a general staff function. Naturally, for most companies, there is a balance between input and output. The injection of a general staff function that is beyond the scale of the enterprise cannot be supported. Usually, to request help from external lawyers and foreign lawyers for direct and indirect services requires considerable expense. However, it is not a wise policy to hold down such expenses more than necessary. Companies should consider what is the injection of legal affairs costs that suits their size and importance.

7.2 EXAMPLE FRAMEWORK OF AGREEMENT COVERING TECHNOLOGY TRANSFER

7.2.1 The States of Technology Transfer

The range and state of technology transfer is of an infinite variety. In the following sections the forms of agreements within a relatively large transfer of technology projects are examined through an example. The parties of this example are listed next, for convenience, and assigned code names (AAA Company, etc.).

There are several transferring parties in this technology transfer example. They are:

- The leading company carrying out the technology transfer (AAA Company). It owns and itself uses the main part of the relevant technology. While acting as the bargaining party, it plans itself to become a party to the agreement.

- A company related to or controlled by AAA Company (aaa Company). It shares part of the technology being transferred.
- An engineering company (BBB Company).
- A construction company (general contractor) (CCC Company).
- A general trading company (DDD Company).

There are two receiving parties in this technology transfer example. They are:

- The main company receiving the technology transfer (XXX Company). It enters into bargaining for the transfer with the ambition to receive the technology, and also perhaps to implement it itself or in a related company controlled by it.
- A company related to or under control of XXX Company (xxx Company).

7.2.2 The Basic Agreement

First, the basic agreement should be called by its name: basic agreement. The basic agreement prescribes the important basic articles of the technology transfer in question. The parties to the agreement are most likely to be AAA Company and XXX Company. Usually, in the agreement preamble, the background of the agreement and the intention of each party concerned is summarized and specified. In the agreement text, in all states of technology transfer, it is specified, on the one hand, what stage of technology transfer will AAA Company carry out and when and how and, on the other hand, what kind of consideration, when and in what form, will XXX Company give or pay. In the case that AAA Company and XXX Company do not become subjected to rights and obligations directly, reference should be made to this in the basic agreement. The obligations that each should hold in connection with this should be specified.

Among the articles that are usually included in the basic agreement, attention should particularly be paid by the transferring side to guarantees and damage compensation in relation to the technology. In guaranteeing the level of performance to be expected when the receiving side starts to use the transferred technology, the transferring side must determine what is achievable, based on its own track record, taking into account unfavorable conditions that may occur in the circumstance of the receiving sides. (For example in chemical plants there are many examples where production capacity should be guaranteed in quantitative terms.) Further points about guarantees relate to the implementation stages of an agreement. In cases of default on an obligation, such as failed performance, late performance, and defective performance, there will be situations when the transferring side would have to compensate the other side.

In these cases, from the viewpoint of legal affairs, the area that should be focused on is the range of damages. It is necessary to specify clearly that damages should be limited to direct damages (for example, in the case of a technical malfunction, repair costs, replacement costs, and the costs of materials wasted as a result of it) and should not include consequential damages (for example, the loss of profits from lost production). If consequential damages are included, the discussion with the receiving side becomes complicated. There is a high possibility that the amount of damages might be bigger than expected.

7.2.3 The Technological License Agreement

Next is the technological license agreement. The parties to the agreement are AAA Company and XXX Company or xxx Company. The articles normally included are shown below:

Objects and range of the license: A specification is made to clarify whether the license is only for a patent or applied-for patent, or whether it also includes technological information (technical data, know-how, etc.); also, what are its purposes of use?

State of the license: Whether it is an exclusive or nonexclusive license. There is a further possibility to allow sublicensing.

Range of technological information and the form of transfer disclosure: The targeted range of technological information is specified clearly, as it can take many forms, such as drawings, manuals, and standards. What form and when transfer is conveyed are specified. The language used for transfer is specified as well.

Technical assistance and training: One of the ways to transfer technological information is for AAA Company or aaa Company to send training staff to XXX Company or xxx Company. Another is for the technical and operation staff from XXX Company or xxx Company to be accepted into the factory of AAA Company or aaa Company for training. Further details on the agreements concerned with these may be found in the Sections 7.2.4 and 7.2.5 on the technical staff dispatch agreement and technical and operational staff training agreement.

Technological license guarantee: Generally, granting the technological license (patent or patent application) does not guarantee anything beyond what is already guaranteed under the basic agreement.

Consideration for license: In general, this article deals with whether payments are made by one side to the other by a deposit or running royalties. In the case of a deposit, lump-sum payment, or payments by installment, when payment should be made and in what currency are described. In the case of royalties, the levy period is specified. Important points are whether the royalties are on sales or manufacturing (mostly they are on sales), whether they are at a fixed rate or of a fixed amount (most cases are fixed rate, although this depends on the industry), on what the royalties levy is based (in the case of fixed rate, gross sales or net sales value), and whether or not a minimum royalty is imposed (in the case of an exclusive license). Furthermore, in the case of collecting royalties, regular receipt of a royalties report is important. It is also important to reserve the right to inspect a licensee's accounts, using an independent accounting firm, if there are reasons to suspect irregularities in the contents.

In addition, for both deposits and royalties, if tax (withholding tax) is charged by the local authorities, proof (for presentation to the Japanese tax authority) should be made available, clarifying both the tax rate and the amount paid.

Confidentiality: The worth of technological information can be maintained only if and to the extent that such information will be treated and kept secret and confidential. It is essential for the transferring party to receive written commitment from the receiving party to keep secret and confidential the technological information to be disclosed under the technological license agreement. First, it is important that the range of maintaining confidentiality should be made clear. Besides written documents, it should be ensured that disclosures by oral means and images shall be included in this.

Next, the period of time is specified over which confidentiality is to be maintained. Also, it is usual to exclude some parts of the disclosed technological information from the recipient's obligation to maintain confidentiality if there are logical reasons and clear evidence for doing so. There are many discussions of the types of case that are excluded, though generally they are limited to the following:

- If the information is publicly known at the time of disclosure.
- If the information is already known by the licensee at the time of disclosure and this can be proved with documentary evidence.
- If the information becomes publicly known after disclosure without any breach of the agreement.
- If the licensee can prove disclosure of the same information from a third party without the obligation of confidentiality.

In addition to the confidentiality provision of the above-mentioned technological information, provision against other purposeful use of the technological information is usually inserted. Since the license for the technological information is granted for an explained basic purpose, it is not allowed to be used by the licensee for purposes beyond that. If it is to be used for other purposes, the range of the license should be widened in the agreement (according to determined criteria).

7.2.4 The Technical Staff Dispatch Agreement

Next is the technical staff dispatch agreement. The parties to the agreement are AAA Company or aaa Company and XXX Company or xxx Company. The articles included are shown in the items below:

Dispatch of staff and location: Details are specified, such as what type of technical staff and the purposes of training, their number, the number of times, and to where they are dispatched.

Specified period: The lengths of time and intervals between dispatch are specified.

Compensation for technical staff dispatched: As the dispatched technical staff leave their home country and move overseas, it is impossible for them to continue their local work. In this case, there is usually a one-off payment to them of a fixed fee called an absence fee. In addition, apart from this lump sum, it is common that they receive an amount of money calculated as a part of the overseas business trip daily allowance. (On the taxation of this, refer to "consideration for license" under Section 7.2.3.)

Maintaining confidentiality: Attention is paid to referring to and being consistent with the article on confidentiality in the technological license agreement.

Other criteria: Necessary information is specified, such as round-trip airfare cost and cost of staying at a local hotel, the transportation facilities to and from the workplace, and response to illness and injury and temporary return home in urgent cases (family celebration or bereavement leave).

7.2.5 The Technical and Operation Staff Training Agreement

Next, in the case of the technical and operation staff training agreement, the operative parties, as in the previous technical staff dispatch agreement, are AAA Company or aaa Company and XXX Company or xxx Company. The articles normally included are as shown below:

Staff accepted and location: Details are specified of the types of staff accepted, the number of staff accepted, to where the staff are accepted, and for what purposes they are accepted.

Acceptance period: The lengths of time and intervals between dispatch are specified.

Compensation for accepting technical staff: Generally, payment of a fixed amount in a lump sum is made when accepting technical staff and carrying out training. (For taxation aspects, see "consideration for license" under Section 7.2.3.)

Maintaining confidentiality: Attention is paid to consistency with the article on maintaining confidentiality in the technological license agreement. When the contents of training cover subtleties of technological information, there are cases where the staff themselves take a written oath.

Other criteria: Necessary information is specified, such as round-trip airfare cost and cost of staying at a local hotel (generally the dispatch side carries the cost; there are cases where the receiving company's dormitory is used). In addition, references shall be made in this agreement to the internal rules to be observed by the trained staff in the receiving party's facilities and to the response to illness and injury of the trained staff.

7.2.6 The Engineering Agreement

The engineering agreement determines the engineering design package that is suited to each case (the layout, process plans, material balances, machinery specifications, measuring equipment list, electric equipment list, etc.). If the receiving side of the technology transfer wishes only to copy the transferring side's plant, there is no need for new engineering. But in other cases, it is necessary to adapt engineering to be suitable for the receiving side. There are times when the transferring side concerned has the engineering ability for this, but in cases when it does not, another company that has such abilities should be included. Hence, the parties to the engineering agreement are AAA Company or BBB Company with XXX Company or xxx Company. The articles of the engineering agreement have varied contents according to the relevant professional field, so they are not reviewed here.

7.2.7 The Plant Construction Agreement

As the next stage after engineering, it is quite common for the side receiving the technology transfer to place orders for the plant construction with the transferring side itself. In Japan's case, there are instances when the business of plant construction is undertaken by a construction company (a so-called general contractor, CCC Company), or in some cases, the engineering (BBB) company also is involved in plant building. From this point of view, the parties involved in the plant construction agreement are BBB Company or CCC Company and XXX Company or xxx Company. The articles of the plant construction agreement have varied contents according to the professional field, so they are not reviewed here.

7.2.8 The Machinery Procurement Agreement

It is natural for the side that receives technology transfer to want to use the machinery that is actually used by the transferring side. Often orders for that machinery result. Even if the parties involved in the technology transfer can handle the ordering of the machinery from its maker as well as management of the delivery at the appointed time, it is appropriate for a general trading company (DDD Company) to handle the supply of machinery to overseas destinations. Such a company has overall abilities in various businesses, such as packing, storage (while waiting for loading), transportation, and customs clearance. Therefore, in many cases, the machinery procurement agreement becomes two agreements, a domestic one and a foreign one. The parties to the agreement in the domestic case are AAA Company and DDD Company, and in the overseas case are DDD Company and XXX Company or xxx Company. The articles of the agreement have varied professional details, and hence are not reviewed here.

7.3 COMMON POINTS TO NOTE IN THE VARIOUS AGREEMENTS' LEGAL AFFAIRS ARTICLES

This section focuses on the legal affairs phrases that commonly appear in the articles of the various agreements of Section 7.2. Although nonprofessionals may find it difficult to understand the many professional terms that are used, attention should be paid to them. They determine what will be carried out after concluding an agreement. It is hugely important that they are properly understood before the agreement is finalized. Particularly people who are not professionals in legal affairs sometimes mistakenly see the topics to be mentioned in this section as light and insignificant. There are many cases of not much attention being paid to these topics and a big amount of money being paid later. At the stage of starting a technology transfer agreement negotiation, generally the transferring side, that holds the technology, has a strong hand. It should not make easy concessions at this stage, and it should firmly press its own point of view. Detailed explanations of the topics follow.

7.3.1 Party to the Agreement

A party to the agreement is in formal terms one (an individual or a corporate body) who signed the agreement. Beyond formal terms, in reality (or substantial terms) a party to the agreement is one who receives rights from the agreement, undertakes obligations, or promises to produce other legal effects. It is necessary for the party to the agreement in reality to possess legal capacity in order to make its carrying out the agreement effective. On beginning negotiations with the other party concerned, problems are not expected with respect to legal capacity when the other party is similar to a Japanese incorporated company. However, when the other partner is a company with a special form, partnership, association, cartel, etc., it is advisable to confirm what is its position with a local lawyer.

7.3.2 Signer to the Agreement

In cases when the substantial person concerned is a party to the agreement, it is obvious that the signature of the person is needed. In the case of a corporation like a company, there is a question of who should sign the agreement on behalf of the company in order that the company may receive rights and effectively undertake obligations under the agreement. In the case of a Japanese joint stock corporation (company), the company has one or more representative directors who can represent the company. Generally there are also other executive or managing directors in the company. When they sign the agreement on behalf of their company, covering the matters of which they are in charge in the company, the rights and obligations will be construed so as to be attributable to the company. In the case of companies overseas, Japanese knowledge cannot be used, as the legal system may differ there. The views of people who well know the legal system of those companies should be referred to. Confirmation should be obtained from local lawyers as the need arises.

7.3.3 Effective Period

It is easy to understand that there is not just one agreement period (from the start to the end time) in a signed and concluded agreement. An end time exists for each of the individual rights and obligations that are specified in the agreement. At every end time when the whole of the corresponding rights and obligations become null and void, the agreement finishes its mission and loses its significance.

7.3.4 Agreement Transfer (Assignment)

In most cases, the technology transferred is an important technology to the party transferring it. After the disclosure of the technology under agreement, if the technology transferred is shown to a third party without the knowledge of the transferring side, the transferring side concerned would suffer unexpected damage. This is prevented from happening through the maintaining confidentiality article in the technology licensing agreement. However, as the whole agreement containing the maintaining confidentiality article could be transferred to a third party, it is common sense to establish as part of the agreement a condition to the effect of "written consent is needed from the other party concerned prior to the transfer of the agreement."

But even flourishing the "need for prior written consent from the other party concerned" is of no avail if substantial parts of the business that received the technology transfer are transferred to or merged with another company. Hence, there are many cases where the above-written phrase, "written consent is needed prior to," is followed by the phrase "this limitation is void in the case of transfer of substantial portions of a business or a merger." This means that once the form of transfer of substantial portions of a business is set, it is difficult for the other party to avoid an early confrontation with the unexpected party. Also, in the case when the other party concerned is similar to a Japanese joint stock corporation (company), it cannot be denied that there is a possibility of change of ownership in the form of stock transfers.

This problem involves the local legal system. Any party who is particularly nervous about it should discuss the point with local lawyers and specialists. It is necessary to consider the lawful maximum countermeasure permitted in the country.

7.3.5 Governing Law

The governing law of the agreement interprets the agreement. It is these laws and ordinances that apply to operating the agreement. The laws of which country the agreement should be based on should be determined. If this is neglected, should problems occur, much time and hard work would be wasted on determining how the agreement should be interpreted, based on which country's laws.

From the point of view of an outcome, it is the governing law, the laws and ordinances, of its own country that should be the deciding law for the party carrying out the technology transfer. Concession should not be made imprudently on this point. This is because in the case of a possible dispute in the future, it would be a great advantage to fight on home ground with familiar laws and ordinances.

7.3.6 Controlling Text

With the exception of specific types of agreements that are not valid without complying with certain predetermined forms, generally it may be thought that an oral agreement is sufficient. But, without any written evidence, it may be difficult for one party to oppose the other party when the other party, on purpose or unconsciously, changes its previous intention or the applicable terms and conditions. Therefore, the contents of an agreement are written down as evidence. The resulting text, signed by both parties, is the controlling text.

The original text should be secured safely. It is enough if each of the parties has a copy and uses that for daily work. In the case when the original is in English, although sometimes a translated version is created and a signature is sought, it is vital not to put a signature to the translated version by any means. It might become confusing in the future.

7.3.7 Entire Agreement

It is common for negotiation of an agreement to take place over a fixed period by many means (face-to-face and video meetings, by phone, fax, email, letter, etc.). In addition to the first draft agreement submitted, generally for the convenience of further negotiations by both parties, a series of revised draft agreements may be prepared. They add, delete, or amend such terms and conditions as both parties have agreed by that time. Through such negotiations, the final draft agreement will be prepared for signatures by both parties. The concept of the entire agreement is that all the terms and conditions that both parties have agreed to so far are involved and incorporated in the final agreement, and that it is written into the final agreement that it supersedes earlier draft agreements. This way of thinking is excellent in that it makes clear to both parties concerned what is intended to be implemented so that both parties can avoid possible future misunderstanding. As a particular point, to avoid later regrets, it is necessary to take care that agreed points, realized as being important at the time of their initial negotiation, are not accidentally lost. As a final point, because the entire agreement replaces and makes void the earlier draft agreements, some people think that a more descriptive term for the entire agreement is previous voiding and final agreement.

7.3.8 Supplement to or Amendment of Agreement

Generally an oral agreement is effective as described in Section 7.3.6, but it is common today that agreements are prepared in written form and signed by relevant parties. Various types of agreements will appear in technology transfer, as explained in Sections 7.2.2 to 7.2.8, and those agreements should have the provisions saying "any supplement to or amendment of the agreement shall not be effective unless such supplement or amendment is made in written form signed by the respective parties." After an agreement is concluded, if new situations occur, or a plan goes wrong, or there are changes to both parties involved in the agreement, both parties should discuss it. When both parties agree to a certain supplement or amendment, both parties should prepare a supplemental agreement or amendment agreement. It should be signed by both parties in accordance with the provisions of the agreement as mentioned above.

After concluding an agreement, working group(s) may be organized by both parties to smooth the path of implementing the agreement. During the course of activities of the working group(s), there may be a case where some documents will be prepared for confirmation or mutual better understanding and be signed by a representative member of each side. Since these documents may affect the supplement to or amendment of the agreement in the future, careful attention should be paid. Although one would not expect the members of the working group(s) to exceed the authority lodged in it, it is advisable to seek the opinion of a specialist in judicial affairs in advance as to what are any legal implications of such signed documents.

7.3.9 Force Majeure

It is rare after concluding an agreement for an uncontrollable situation, such as a natural disaster, war, or explosion, to occur during the agreement's implementation. However, if it does, the implementation of the agreement may be totally or partially affected. In most countries civil laws or civil courts determine certain relief or a basis for claims. However, their contents and level may vary. Therefore, in cases of emergency, what would certainly be received in what particular situation, rather than leaving things uncertain, should be considered. It is wise to consider what is the least that can be done, clarify as many points as possible, and incorporate them in the agreement.

7.3.10 Termination of Agreement

After concluding an agreement, the other party concerned may for some reason neglect execution of an obligation or may perform an act that breaks the clauses of the agreement.

When such a situation occurs in the case, for example, of a one-time sales agreement, the cancellation of agreement (terminating the agreement retroactively to the beginning) sets out the following steps: immediately canceling the agreement, returning the goods and money from sales, and returning to the state before agreement (the original state), besides ending with compensation for the damages incurred.

However, in the case of an agreement concerning technology transfer, it may take a long time to accomplish the whole performance of the agreement, or once transfer has started, it is not possible to return to the original state (for example, when technological information has been disclosed). In this case, instead of a cancellation of agreement concerned with returning to the original state, it is common to set out the agreement in the form of a termination of agreement to come into force at some time in the future. Even if there is nonexecution by the other side, the agreement is not terminated immediately, but a fixed notice period is set. Only if a requested execution of obligation is not carried out within the set period is the agreement terminated at the end of such a set period and the agreement loses its effectiveness for the future. Needless to say, it is in many cases accompanied by compensation.

7.3.11 Settlement of Disputes

Even if an agreement is duly concluded and the actual business activity based on it starts, it does not necessarily progress harmoniously. It is common for problems, both small and large, to occur at the various steps of implementation of the agreement. There are many kinds of possible problems. Three examples are the occurrence of a situation not expected by the agreement, different interpretations of the agreement clauses, and one party neglecting its obligations. It is most preferable that disputes between the parties concerned are harmoniously solved by discussion between them. However, it is not as easy as said, because in reality, there are many unresolvable cases. Hence, these should be prepared for in the agreement, assuming a worst-case situation.

Generally, the applicable terms here are litigation, conciliation, and arbitration. Although a lawsuit (litigation) has the merit of clarifying what is black and what is white, there is a risk of a contrary ruling. In addition, generally it requires a long period to settle a dispute in this way. Furthermore, the expense of a lawsuit, including legal fees, can turn into a large sum. Although conciliation can be straightforward, particularly when independent conciliation organizations exist in a country, there is a risk that no conclusion is reached if the parties concerned do not agree with the conciliation plan created by the conciliator.

In contrast, arbitration is a process whereby an application is submitted to a fixed arbitration body, an arbitrator is selected, the arbitrator hears opinions from both sides concerned, and an arbitration award is handed over. The point about arbitration is that it has the power to restrain both parties, and also the power of execution based on each party's civil law. However, the power of execution based on civil law comes from an international treaty (1958—Convention on the Recognition and Enforcement of Foreign Arbitral Awards (New York)). One should be careful that this treaty applies to the countries of the parties.

Generally, arbitration takes a shorter time and costs less compared to a lawsuit. Therefore, there are many cases of international agreements, such as technology transfer agreements, adopting arbitration to settle disputes. Nevertheless, there are various points, considered next, to which attention should be paid.

7.3.12 Arbitration

When adopting arbitration to settle a dispute, the following points should be taken into consideration:

Place where arbitration proceedings are carried out: The party that transfers the technology should gain implementation of arbitration in its own country at any cost, taking advantage of the strong position that it has at the time of negotiating the agreement.

For the transferring side, the worst legal affairs situation that can happen is accepting both the applicable law (Section 7.3.5) and the place of arbitration to be that of the other party's country. However, the other party may insist that its own country be the arbitration place and may not give way. In that case, between A and B, the principle of reciprocity may be adopted. If A applies for arbitration, the arbitration place would be in B's country, and vice versa.

When one party refuses to accept such reciprocity, there are also cases, although rare, where the place of arbitration is set in a third country, neutral for both parties concerned. If it is a country near to the country of a party concerned, it is still fine, but in the case of a distant country with an unfamiliar language, there will be unexpected expenses, including the translation expense of a presentation document, etc.

Arbitration rules to adopt: An arbitration proceeds in some manner determined by an internationally recognized arbitral organization. As the arbitral organization of each country (for example, Japan Commercial Arbitration Association (JCAA), International Chamber of Commerce (ICC) (headquarters in France), and American Arbitration Association (AAA)) has its own arbitration rules, agreement should be made beforehand on the organization whose arbitration rules are to be followed. However, this does not necessarily link to the place where the arbitration proceedings are carried out.

Arbitrators: Usually the method of selection of an arbitrator is determined by the arbitration rules described above. Although there are normally more than one arbitrator (an odd number), as the number increases, cost will increase in proportion. In the case where there are three arbitrators, usually a neutral arbitrator would be nominated in addition to the two nominated by the parties.

Language used: Independently of the arbitration place and the arbitration rules adopted, the language used for arbitration is clarified (for oral explanation, presentation and creation of data, etc.).

Arbitration expenses: In many cases, the expenses of arbitration are shared between the parties concerned (naturally, the legal fees of each party concerned, arising from defending its position, must be taken up by the party itself).

DISCUSSION QUESTIONS

1. The conclusion of many agreements is necessary as part of transferring technology to different countries. A final outcome of the agreements should be that both the transferring and receiving sides are satisfied. Agreements take different forms according to the field of technology that is involved. Assuming that you are in charge of drawing up agreements for the transferring side in one of the technical fields of chemical processing, mechanical parts, electric products, or any field with which you are familiar, make a list of the relevant agreements and explain what you would include in them.

2. Sometimes, even though technology transfer has taken place in accordance with the agreements, the resulting product quality is not satisfied. In such cases, various countermeasures can be taken. What actions do you take as a person in charge of the agreement? Describe your ideas about countermeasures. Consider your answers from the points of view of both the transferring and receiving sides.
3. Explain what are the particular points that a receiving side should consider when receiving technology from another country.

8

Technology Transfer from Participants' Viewpoints

In recent years, globalization and borderless industrial structures have grown to cover the whole world. International activities such as overseas production, mergers and acquisitions, and global alliances are occurring more and more frequently.

Technology starts by creation. It then spreads in two ways. One is by transmission, for example between generations (senior to junior). The other is by transfer, for example between companies and between nations. If some core technology is created, various opportunities will open up. The industry concerned will develop and grow as the technology is handed down and transferred. If technology is held back and not transferred, a competing technology will appear someday. It will be difficult for a company to maintain a predominant position even though short-term profits may be safeguarded. Therefore, technology transfer can be said to be an unavoidable phenomenon.

When the main driver of technology transfer overseas was reduction of labor costs, technology transfer was strongly associated (Figure 8.1) with negative images, such as domestic contraction, outflow of technology and a brain drain, fall in global competitiveness, and criticism from overseas (exploitation/discrimination/friction). But with China's and Japan's manufacturing activities in recent years as an example, it is now seen to be necessary to tackle overseas technology transfer positively, as a way to solve problems, and relate it to the development of the local area. As represented in Figure 8.1, technology transfer has taken on a positive new image, one in which it is fundamental to gain profit with overseas countries from the development of coexistence and mutual prosperity. Win-win solutions are created by collaborations between technologically established Japan and overseas countries.

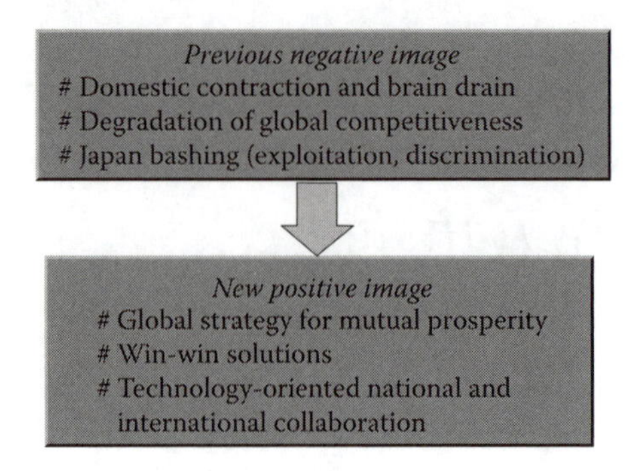

FIGURE 8.1
Paradigm shift in technology transfer.

There are risks in overseas technology transfer, and there are many reported examples of failures. Many failures arise from poor initial assessments of the situation. However, if the transferring and receiving sides can share profits, it will become a valuable and successful experience for both.

This chapter first reviews the background of technology transfer needs. It then reports the results of a survey of advanced technology transfers to Thailand, China, and Indonesia, carried out by members of Hiroshima University between 2005 and 2007, funded by Japan's Ministry of Economy, Trade and Industry (METI). Table 8.1 lists the companies visited and what is their business area. Table 8.2 gives information on their size, as well as when they were visited.

Issues for the receiving side are particularly focused on and analyzed. The issues, from both sides' points of view, are divided into individuals' issues and company, education, and local government issues. Ways to tackle and solve problems are put forward.

Asian nations take on board the techniques from new technology transfer. Both the transferring and receiving sides expect to achieve global economic development for their mutual benefits.

8.1 BACKGROUND OF TECHNOLOGY TRANSFER

8.1.1 The Scope of This Chapter

There are not-for-profit enterprises, carried out, for example, by state and public institutions, and enterprises carried out for profit, usually by private companies. Although technology transfer is usually associated with the

TABLE 8.1

Companies and Organizations, with Their Country and Activity, Visited for the Survey

Companies and Other Organizations	Country	Business or Other Activity
Satake Thailand Co., Ltd.	Thailand	Rice mill machines, etc.
Nishikawa Tachaplalert Rubber Co., Ltd.	Thailand	Rubber products for auto door seals
Chulalongkorn University	Thailand	Technology transfer
Satake Manufacturing (Suzhou) Co., Ltd.	China	Rice mill machines, etc.
Shanghai Nishikawa Sealing System Co., Ltd.	China	Rubber products for auto door seals
Changzhou Baoling Heavy Industries Co., Ltd.	China	Mill equipment, etc.
Changzhou Baoling Heavy and Industrial Machinery Co., Ltd.	China	Iron manufacturing facility design
Shiseido Liyuan Cosmetics Co., Ltd.	China	Cosmetics, toiletry goods
Takara Brewery Food Products Co., Ltd.	China	Brand sake, distilled spirit
Wacoal Fashion Ware Co., Ltd.	China	Women's underwear
Japanese Chamber of Commerce and Industry in China	China	General
Institut Teknologi Bandung	Indonesia	Technology transfer
PT.Kasen	Indonesia	Spinneret nozzles, auto parts
PT.Sakura Java Indonesia	Indonesia	Auto and motorcycle parts
PT.Kyowa Indonesia	Indonesia	Auto parts
PT.Indonesia Epson	Indonesia	Printers
Manufacturing Industry Development Center	Indonesia	General
PT.Ayawinsa Dinamika Karawng	Indonesia	Auto parts
PT.Kyoda Mas Mulia	Indonesia	Auto parts, rubber materials

latter, this chapter will consider more broadly how both private and public enterprises, both on the transferring and receiving sides, can cooperate to achieve mutual win-win solutions.

8.1.2 Japan's Needs for Technology Transfer

The structure of Japanese-style economic development started to break down in the 1980s. The features of Japanese business that drove the development included good quality labor, a lifelong employment system that resulted in staff loyalty and stable management, the special nature of

TABLE 8.2

Year of Visit and Company/Organization Financial and Employee Data

Companies and Other Organizations	Year of Visit	Share Capital (j, joint; g, group; i, independent)	Capital Fund (billion yen)	Employees (year of visit)
Satake Thailand Co., Ltd.	2005	j	0	≈210
Nishikawa Tachaplalert Rubber Co., Ltd.	2005	j	2	≈600
Chulalongkorn University	2005	—	—	—
Satake Manufacturing (Suzhou) Co., Ltd.	2006	i	1	≈280
Shanghai Nishikawa Sealing System Co., Ltd.	2006	i	3	≈900
Changzhou Baoling Heavy Industries Co., Ltd.	2006	j	4	≈1920
Changzhou Baoling Heavy and Industrial Machinery Co., Ltd.	2006	j	0	≈30
Shiseido Liyuan Cosmetics Co., Ltd.	2006	j	1	≈3280
Takara Brewery Food Products Co., Ltd.	2006	j	2	≈110
Wacoal Fashion Ware Co., Ltd.	2006	i	2	≈1100
Japanese Chamber of Commerce and Industry in China	2006	—	—	—
Institut Teknologi Bandung	2006	—	—	—
PT.Kasen	2007	i	0	≈260
PT.Sakura Java Indonesia	2007	i	0	≈750
PT.Kyowa Indonesia	2007	i	0	≈610
PT.Indonesia Epson	2007	i	3	≈8400
Manufacturing Industry Development Center	2007	—	—	—
PT.Ayawinsa Dinamika Karawng	2007	i	?	≈1000
PT.Kyoda Mas Mulia	2007	g	?	≈250

Japanese culture, protective policies whereby the government and companies cooperated, and a favorable accounting system. These are no longer believed to be weapons that can win straight victories in the present highly competitive world.

Japanese companies need an international strategy for survival. It is inevitable that they decide to manufacture overseas or transfer technology. But seeking the labor cost reductions that have been typical is not enough. They need to develop new concepts of coexistence and mutual prosperity while maintaining a competitive cost and performance.

8.1.3 Asian Nations' Needs for Technology Transfer

There have also been remarkable changes among the Asian nations. Belief in the market economy and a free economy has become firmly established after the collapse of the Cold War structure. The environment is established in which the need for technology from developed countries is actively accepted.

The first stage of change for Asian nations was becoming the global production base for key industries, such as fibers, iron and steel, and shipbuilding. The need for high technology and the modernization of technology has emerged as the second stage, with the aim of transforming to an industrialized country.

8.2 NEW TECHNOLOGY TRANSFER—ISSUES THAT SHOULD BE TACKLED

Japan started to transfer technology to Asian nations in the 1960s. Examples were in the textile, shipbuilding and steel, automobile, food, and electrical and electronic industries. Concern about the decline of domestic industries spread during the early period as a result of overemphasizing the goal of labor cost reduction, even though special and advanced technology stayed in the home country. It needs to be realized that it is next to impossible to win outright victories in an international competition. Compromises are necessary. In recent years, the balance of advantages has become an important subject in considering manufacturing activities between Japan and China or India.

This need to balance makes it vital for those actively involved in technology transfer to grasp the importance of the following points: coexistence and mutual prosperity, building of win-win solutions, and building a global strategy that aims at fusion between technologically established Japan and overseas countries. At the same time that, for example, small and medium-sized enterprises approach and set up organizations in Asian countries, they must also continue to seek new domestic innovation.

To help understand these points, it is useful to investigate advanced examples of technology transfer that have taken place in recent years, to study and analyze any underlying issues and problems, and to propose ways forward for future successes.

8.3 A TECHNOLOGY TRANSFER SURVEY

8.3.1 Purpose of the Investigation

There are many previous reports that consider problems of overseas technology transfer from the point of view of the transferring side. The study reported here investigates problems and issues from the receiving side's point of view as well as from the transferring side's. Current situations in Thailand, Indonesia, and China are taken as examples. Various solution policies and new methods of supporting technology transfer are suggested in the light of the study.

8.3.2 Survey Outline

Tables 8.1 and 8.2, mentioned earlier, list the countries and organizations visited for the survey, and the year of visiting. General questions were asked about the background to the technology transfer, the scale of the business, the current transfer situation, as well as what were the specific issues and problems. Wherever possible, both employees from the receiving side and managers from the transferring side were questioned. In Thailand, additional general information on receiving side views was available in the form of published university and public institution documents.

8.4 RESULTS FROM THE SURVEY

The many opinions and problems mentioned in questioning the companies, public bodies, and universities were analyzed. Here they are grouped by issues mentioned by the receiving sides and the transferring sides and summarized below. It is interesting how the positions and viewpoints of both groups differ from each other. Some points were country specific. These are summarized too.

8.4.1 Issues as Seen by Receiving Sides

- The transferring side begrudges the transfer of technology.
- Transfer often involves defects and failures.
- Differences between the specifications and the inspected drawings and parts are also causes of defects.
- There are insufficient skills (on both the transferring and the receiving side).
- Information is not sufficiently disclosed; the transferring side inserted limitations into the agreement.
- The technology is outdated. There is little advanced technology.
- There is little spin-off benefit to allied industries, universities, and research institutes.
- It is necessary to be familiar with Japanese-style management.
- The transferring side wants to be in charge of management as well.
- The Japanese have higher salaries; salary improvements are desired.
- There are barriers in communication (problems of mutual understanding).

8.4.2 Issues as Seen by Transferring Sides

- Workers require on-the-job training in manufacturing skills.
- Training of key persons is important, and sending them to Japan is expensive.
- There is no tradition of reporting/communicating/consulting; a lot of attention must be paid to production control.
- There are many cases of job hopping (job changing), making imitation goods, and illegal selling of drawings.

- Research and development is still difficult to be carried out locally (lack of talented people, funds and technical protection).
- Differences in national traits make it hard to become familiar with local administrations and management ways.
- There are barriers to communication (explanations take three times longer than is normally necessary, etc.).

The above responses indicate many of the typical issues. It might appear that techniques of technology transfer are not making progress. But in reality the technical contents of transfers are developing every year, and new problems are gathering at the same time. Even though many well-known problems, conventionally regarded as arising from differences in organizations, cultures, and customs, were pointed out, such as to do with personnel management, salary organization, and executive personnel affairs, these days remarkable improvements have been made. On the other hand, arguments continue around the dilemma of advanced technology transfer or domestic technology protection.

8.4.3 Country-Specific Issues

In Thailand:

- There are many occurrences of technical leaks and job hopping.
- Infrastructure is set up, and domestic industry is developing as well.
- The receiving side is well known for transferring technology imperfectly.
- Communication is a challenge. Learning the Thai language is difficult, so English becomes the official language. However, English is not generally understood by workers.

In China:

- The workers' determination and spirit of self-advancement are high. Education packages are welcomed.
- The salary system, employment rules, and maintenance of personnel management are extremely important.
- There are many networks for attracting enterprises, but it is necessary to be aware of differences between the social system and the Japanese one.
- There is little communication in English because of the common use of Japanese and Chinese.

- Measures besides rights of patent are necessary for technical protection.

In Indonesia:

- Both the government and people strongly expect friendly relations with Japan.
- It is necessary that rules and systems fit in with Islamic culture.
- There is a labor law to limit the period of employment of workers under fixed-term contract. Employment planning is difficult.
- There are many cases where the local parts industry cannot manufacture to the standard required by the Japanese-affiliated company (guidance from the Japanese side is expected).

8.5 ROAD MAP FOR RESOLVING PROBLEMS

The survey has identified a wide range of problems. Some may be resolved by individuals, while others are mainly the responsibility of companies. Yet others may be overcome by university and institutional involvement. Others again need to be tackled at the local authority and national levels. This section first repeats what are the needs of the transferring and receiving sides. Then it presents recommendations and proposals for how technology transfer may be improved, and mapped on to the personal, company, educational, local, and national divisions just mentioned.

8.5.1 Differences between the Transferring and Receiving Sides

Transferring and receiving sides participate in technology transfer for fundamentally different reasons.

Purposes for the receiving side are:

- To completely absorb the technology over time (including background and neighboring technologies).
- To influence its related industries and organizations, to help to develop the economy.
- To absorb advanced technology.

Purposes for the transferring side are:

- To minimize the costs and time needed for transfer.
- To acquire the value from technology transfer at an early stage.
- To avoid disclosing advanced technology related to core competences.
- (Influencing other industries is not of interest).

Since technology transfer is handled as a business, it is natural that these differences of purpose give rise to problems. Beyond recognizing the differences between the two parties, it is vital that efforts are made to reduce the various sources of conflict.

8.5.2 Issues Arising at the Individual Level

The person in charge of technology transfer has a wide range of responsibilities. If an unfavorable situation develops, liability will fall not only on the company, but also on the individual. Particularly, it is important to keep an accurate and detailed record in case the receiving side claims that there are faults in a transfer. Issues raised frequently by receiving sides (Section 8.4.2) that may be the responsibility of the person in charge (they may also be for the company to solve) are:

- The transferring side begrudges the transfer of technology.
- Transfer easily becomes second best. It can be defective as well.
- Information is not sufficiently disclosed.

Possible causes are:

8.5.2.1 Cause 1: The Personality of the Individual in Charge

The job of the person in charge of technology transfer is to carry out the transfer with sincerity, passion, and patience based on a spirit of mutual understanding. Explanations and communication of information by the transferring side are seen to be inadequate in many cases. It should be recognized that the receiving side may need something to be explained more than once. People receiving instruction often mention instructors' haughty attitudes. It is vital to be modest and to present technology transfer as a natural development. In addition, it is useful to have an understanding of management of technology (MOT) or related subjects.

If people on the receiving side have anything they do not understand or any other concerns, they should repeatedly ask the transferring side until the point in question is answered. Furthermore, efforts should be made to raise the basic learning levels and basic skills and to improve the general conditions of the receiving side.

8.5.2.2 Cause 2: Not Understanding the Technology Transfer Agreement and Its Range

Things going wrong in technology transfer are often related to the agreement signed in the first place. The person in charge of the technology transfer should aim fully to understand the agreement and its limits well before starting to carry out the transfer. If anything is unclear, it should quickly be clarified with higher management. The outcome should be confirmed in writing to the other party. Serious problems can be caused by a person in charge carelessly making requests or promises outside the range of the agreement.

Further dangers can arise if a person individually tries to make good deficient or unclear areas in the agreement. The jurisdictional limits of an individual and organization are clearly defined. If something beyond that authority is involved, consultation with higher management must take place and advice must be sought promptly.

8.5.2.3 Cause 3: A Language Barrier

A good command of English is required for technology transfer. In some cases there may be a need to learn the local language. The target is not to speak fluently but enough to be understood. The target is to gain the other party's understanding through clear explanations, repeated as often as necessary, and to listen to a partner's remarks patiently in order to understand his questions and problems. This is vital. Referring to the view that explanations take three times longer than is normally necessary, the individual must develop communication skills to overcome that, to make technology transfer smoother.

8.5.2.4 Cause 4: Insufficient Basic Learning and Skills on the Receiving Side; and also

8.5.2.5 Cause 5: Inherent Problems in the Transfer Process

These deficiencies and problems are generally ones that companies and organizations should deal with. When a person in charge faces them, it is important that he brings them to the notice of and follows directions from higher management.

Thus, although technology transfer should be carried out in a spirit of mutual understanding, the person in charge has to work carefully within fixed bounds. In addition, it is important that he keeps a record of actions and meetings in case trouble occurs. From this, what the person is responsible for can clearly be distinguished, and unfounded criticisms are stopped.

8.5.3 Issues Arising at Transferring Company Level

Issues tackled at the company level are most significant. If the technology transfer becomes unfit for its purpose, it will be serious for both sides. The following problems from the receiving side that a company often faces are crucial ones, with the causes afterwards.

- Transfer easily becomes second best. It can be defective as well.
- There are insufficient skills (on both the transferring and receiving sides).
- Information is not sufficiently disclosed; the transferring side inserted limitations into the agreement.
- The technology is outdated. There is little advanced technology.

8.5.3.1 Cause 1: Unclear Agreement Documents and Lack of Mutual Understanding

The success or failure of technology transfer by a company is greatly influenced by the agreement. Therefore, the transferring side must describe the transfer agreement in detail. The receiving side must understand the agreement's contents. Possible problems should be identified and settled in advance.

Since the futures of both the transferring and receiving sides are at stake, it is important that they are both clear about the transfer policy. For example, in the case of transferring mature technology, it is know-how that is important, and its range of disclosure should be made clear. In transferring advanced technology, it is the core technology that is important, and it is its range of disclosure that should be made clear. In addition, it is vital to make sure that both sides will gain from the transfer.

It is very difficult to put right any defects in an agreement or its interpretation after technology has been transferred and operations started if defects and poor running only come to light then. This is even more the case if items outside the range of the agreement have been promised or secured at the discretion of the person in charge. These possibilities must be guarded against.

8.5.3.2 Cause 2: Inadequate Risk Management

Although companies usually carry out feasibility studies as a matter of course, contingency studies (preparations for unexpected situations) tend to be neglected. For example, it is easy to assume that some high-performance technology will function when set up overseas in the same way as it would in Japan. However, because of factors such as buildings, facilities, machinery, parts, raw materials, secondary raw materials, and utilities, it is impossible to transfer everything to do with the technology from Japan. Many items will come from local supplies and resources. These differences accumulate. If any quality or process problems arise, it is usually these differences that are the cause.

Therefore, as part of considering possibilities of malfunctions, it is necessary to agree in advance which party is responsible for what and also procedures for putting things right.

In Asian countries, there are many cases where the receiving side requests all local failings to be fixed by the transferring side. Caution is required here. Appropriate procedures should be set up by mutual agreement.

8.5.3.3 Cause 3: Agreement Documents Not Anticipating All Problems

Unforeseen problems can occur. They can be major ones. The agreement needs then to be revised, in the light of the information from those involved with the problem, with the agreement of all parties. The agreement must be obtained in writing. Criticisms by the receiving side, such as the technology transfer is being interfered with or the information is being restricted, frequently arise from these unforeseen problems around the edges of the agreement. It is most important to make an agreement as clear and all-encompassing as possible.

8.5.3.4 Cause 4: Difficulties in the Management of Technology (MOT)

The targets of technology transfer are specific products and processes. These are identified in the agreement documents in terms of engineering techniques. However, techniques alone cannot guarantee either the product quality or the outcome of a process. A tradition of engineering management and workforce technical skill is required to bring about a normal production activity. Therefore, skills transfer is also a target of technology transfer. However, because a part of skill is the ability to recognize and make decisions about unusual events, there are aspects that cannot be documented.

Companies are responsible for many efforts to improve engineering management and skills as part of technology transfer. It is important that an effective education and training plan is prepared taking into account the basic learning skills and the practical skill level of the receiving side. In many Japanese companies, veteran engineers are sent to the local area to instruct in skills by on-the-job training. Local engineers who become key persons are sent to Japan to master advanced skills.

An important aim of this book is to document and illustrate parts of this tacit knowledge transfer as much as possible, and to turn it into fixed guidance. There is a need for more case studies (such as the survey on which this chapter is based) and testing of the conclusions.

8.5.4 Issues Arising at an Educational Level

Receiving side issues, followed by causes, to do with university and other educational institution affairs are:

- There are insufficient skills (on both the transferring and receiving sides).
- It is necessary to be familiar with Japanese-style management.
- There are barriers in communication (problems of mutual understanding).

8.5.4.1 Cause 1: Insufficient Basic Education

The tradition of skill is an important part of technology transfer. For this reason, on-the-job training exchanges between the local area and Japan, although they are common, may not be enough. There could be a gap between the learner's ability to acquire skill and the instructor's ability to impart it.

Basic skills are acquired as part of basic learning in educational establishments. They are reinforced by exercises and practical work. Generally, the receiving side lacks basic scholarship and basic skills, and in many cases the transferring side does not have an instruction ability in these areas. These failings indirectly lead to poor product quality and manufacturing process problems. Educational institutions, such as universities, are looked to to enrich practical education, by exercises, basic experimental studies, and training.

8.5.4.2 Cause 2: Shortage of Cultural Exchange Education

Although cultural exchanges have prospered in recent years, it cannot be yet said that there is enough mutual understanding between peoples of different countries. Although a Japanese company applies Japanese management methods of reporting, communicating, and consulting when transferring technology, a non-Japanese person may not master this easily. There are also different senses of values from country to country. For example, Japanese people have a culture in which money matters are not mentioned immediately and openly, but people in other Asian countries may ask about salaries or compensation schemes right from the start. This may lead to mutual unpleasantness.

Many foreign students have been accepted into the universities of every country and mutual understanding is increasing. Both Japanese and foreign students take the course on technology transfer from which this book has emerged. The Japanese students are developing the abilities and character required by those on the transferring side. If the foreign students were to become the persons in charge on the receiving side, it could be expected that many of the issues in transferring technology could be solved.

8.5.4.3 Cause 3: A Language Barrier

Although practical language teaching has been encouraged in universities and other educational facilities in recent years, it cannot be said that the results are yet good enough. In particular, it is important to spread English language as a way to communicate. It is needed for the speaking and negotiating abilities required by businesses. At the same time, international mutual understanding is deepened.

8.5.5 Issues Arising at Local and National Levels

Although the companies and individuals directly involved are capable of carrying out the large part of technology transfer themselves without outside help, when it comes to matters to do with a country's business environment, the support of more powerful organizations, local government, and the state is indispensable. The following are issues and causes where government level actions influence transfer.

- There are insufficient skills (on both the transferring and receiving side).
- The technology is outdated. There is little advanced technology.
- There is little spin-off benefit to allied industries, universities, and research institutes.
- The transferring side wants to be in charge of management as well.

8.5.5.1 Cause 1: The Business Environment and Laws of the Receiving Country

A common receiving side criticism is that advanced technology is only rarely transferred. However, if intellectual property protection and laws against imitation goods are not in place and policed in a receiving country, the transfer of advanced technology is difficult there. One expects the following improvements at the local or national government level.

- Strengthening of intellectual property protection.
- Reduced demands on the transferring side by government (reduced requirements for local purchasing, more financial help, state guarantees, more favorable regulations).
- Reasonable requirements on the transferring side (amount and terms of payment, guarantee period).
- Building the support organization that a country needs (grants-in-aid, legal system, information services).

If an environment for technology transfer similar to that in Western countries were to be achieved, it could be imagined that transfer of advanced technology, revealing of detailed know-how, transfer of management to local people, etc., could progress. Currently, measures cited by transferring sides as necessary for defending against the pirating of their advanced technology include keeping control of drawings of important parts, restricting the supply of important parts, and employment contracts restricting key persons' rights to move away.

8.5.5.2 Cause 2: Insufficient National Support

A common criticism by the receiving side is that industry related to the main technology transfer (for example, auto parts) does not grow even if it invests in the required technology. Although to some extent the transferring side can educate local related companies in how to improve themselves, there is a limit to that. There is a need for more systematic support and development, for example, through national industry associations, as in Japan.

8.5.6 Communication and Language Barriers

The issues of communication and language barriers that have been mentioned previously under various headings are returned to here. What are the linguistic needs of technology transfer is considered. The English language is taken as an example, as it is generally the official language for technology transfer.

The need is not necessarily to speak with perfect English grammar. Even when a speaker uses an English word and a listener hears the same word, the two may give different meanings to it, because of their different cultures and customs.

What is important is the linguistic ability to communicate from one to the other. Powers of expression and a passion to be understood is required of the speaker. A high degree of patience and a desire to comprehend what the other is really saying is asked of the listener.

In the survey on which this chapter is based there was some indication of a preference in China and Indonesia for the local language, and not English, to be used. The view was that misunderstandings decrease when the local language or Japanese language is used.

Regardless of the language used, the language abilities required for technology transfer largely depend on the nature of the individuals who are involved. There is a reason for the saying "Technology transfer depends on people, after all."

DISCUSSION QUESTIONS

1. What are common issues often put forward by receiving sides? Give five examples.

2. What are common issues often put forward by transferring sides? Give five examples.
3. If you have experiences of your own in receiving or transferring technology, give your own examples with suggestions for how to improve matters.

9

Overseas Expansion Technology Decision Making

Chapter 6 describes the historical development of Japan's manufacturing overseas expansion. It includes financial and market growth reasons for expansion and company views of future trends. It also touches on decision making concerning where to expand, through describing the planning process by which investment decisions are made. Chapter 8 is about transferring and receiving sides' opinions of problems and difficulties of overseas expansion after the decision to expand has been made. This chapter returns to the question of decision making on where to expand, but from a skills and technology capability point of view.

9.1 OVERSEAS EXPANSION AND THE LEARNING CURVE

9.1.1 A Way of Thinking to Underpin Overseas Expansion

How companies combine their human, physical, financial, and information resources is important to their success. This is true whether they are manufacturing at home or overseas. The biggest problems in overseas expansion are perhaps the human ones. In the home country personnel planning ensures that there is a continuity of human resources. In overseas expansion, in countries without a pool of skilled labor, the human factor starts from zero. Although domestic (Japanese) help can be expected at the start, there comes a time when local engineers and technicians need to be relied upon. An important question is whether or not enough time can be made available for training. This applies both to the training of local engineers and technicians by sending them to Japan and to on-the-job training of the local side by sending engineers and technicians from Japan. In domestic Japan, the mastering of skills is gained through experience over a long period of time. This way of learning cannot be guaranteed overseas. The workforce may be less stable too.

Therefore, let us try to think of overseas expansion in terms of the learning curve introduced in Chapter 2. The assumptions are as follows:

- The same products are to be made at home and overseas.
- Therefore, the level and content of the domestic and overseas technology should be the same.
- Company activities overseas should use local people and resources wherever possible.
- The shorter available training period for local than domestic workers may lead to different results at home and overseas.
- The local workforce may be less stable than the domestic one.

The first three points are the basic ones. There is a view that it is low-priced goods, with reduced functions, that should be manufactured overseas. However, it is essential that the quality of products is the same no matter where they are manufactured. Otherwise, a company's corporate image will suffer. It is this that demands that there should be the same technical level overseas as domestically. In addition, on the point that local people and resources should be used wherever possible, it is only by this means that the company expanding overseas and the local company and area gain. The company expanding overseas gains from reduced labor costs. The local company and area gain from increased employment and an economy developing from the technology transfer. The expectations of all parties are satisfied, with the employment of local engineers as an essential element. Moreover, if manufacturing at the local site is profitable, including transport and any quality costs, it is only right to employ and train the local workforce.

9.1.2 Is the Learning Speed Different Overseas?

Problems in overseas expansion are related to the human factors in the fourth and fifth bullets of Section 9.1.1. The workers employed usually would start from a skill level close to zero. In terms of the learning curve, it is quite normal that Japanese workers and local workers are associated with different unproductive (training) times and rapid learning times (the times L and T in Figure 2.7), and with different standard deviations around the average learning curve. It is a big problem and the main uncertainty in overseas expansion. If the time to attain a certain skill level could be accurately predicted, it would be extremely useful to the drafting of production plans. Obtaining hard data and statistical processing would be essential for this.

One of the factors determining the overall learning time is the level of difficulty of the work, whether it is simple, standard, or skillful work. There should not be a problem in hiring local workers, with no experience, for simple work. Such work needs a shorter preparation compared to general work and requires only a short time to be learned. On the other hand, hiring a worker for skillful work poses a big problem. It is difficult to assess how long a person will take to prepare for and learn the required skills. It may even be that a person does not have the ability ever to acquire the required level of skill, at least within a time following which the investment in the training will pay off. This clearly depends on both the ability of the workforce to learn and the long-term employment stability of the workforce. It is hard to transfer skillful manufacturing that has a long time constant for learning to a foreign country in which there is a high turnover of the workforce, because maintaining the training of high-level engineers and technicians is difficult.

From the point of view of where to expand, the question becomes whether the skill learning speed depends on country or area? To answer this, first it must be considered what elements influence the skill learning speed. From experience of technology transfer, there are three relevant elements:

- The people who do the transferring.
- The technology and skill that is being transferred.
- The people who receive it.

In the case of the people who are transferring, it is necessary that they have transferring know-how and technique. There is a saying: "Even if one is an excellent player, it does not make one an excellent coach." The ability to transfer technology in a limited time is vital. It is particularly difficult if there are communication problems and differences in values between transferrers and receivers. The time factor limit is an important problem that cannot be avoided. In overseas expansion, it is impossible to think, as is the case in Japan, that it takes 5 to 10 years until people are trained. In general, the need is to take the shortest time possible to bring them up to speed and to make profits. The need to train people quickly, and hence the ability to do it, does not exist in domestic Japan, where a large pool of skillful workers already exists.

In the case of the people receiving the transfer, their determination to learn the technology and skill is the most important, followed by their ability to receive it. The idea that being employed by a company automatically means a worker is determined to learn the necessary technology and skills cannot be said to be completely true even in Japan today. The point about determination is particularly important in countries with different cultural values than Japan's.

As for the ability to receive technology and skills, education level is the key issue. In simple work, there may be no need to understand background theories and principles of the work, such as why this should or should not be done. But in work that needs at least a few years of experience for it to be mastered, it is easy to believe that understanding why this should or should not be done would make a big difference to the speed of learning. The learning method when principles and theories are not understood is to repeat and remember the steps taught. But when principles and theories are understood, workers can understand what needs to be done themselves. Depending on the case, they can even add their own planning to it. The speed at which specialized skills are mastered by a company worker would be increased if the required theories and principles were learned before entering the company and if education in the specialized field were available as part of a country's higher education system.

Summarizing the above, the answer to the question whether the skill learning speed depends on country or area itself depends on the following:

- If there is someone able to transfer the technology and skill in the local area.
- If there is enough time for training the workers.
- If there is a determination by the workers to learn the technology and skill.
- If there is a sufficient education level for the technology's underlying theories and principles to be understood.

9.1.3 Decisions to Be Made When Expanding Overseas

It has been considered how differences in type of work (simple, standard, skillful) and the levels of skill of local engineers and technicians influence problems of overseas expansion. Once a decision is made to expand overseas, the need is to start production as soon as possible after investing in the expansion. It is necessary to consider how to obtain the required numbers of engineers and technicians of the right level for the different jobs. It is necessary to answer the following questions:

- How many Japanese workers are needed at the local site and for which jobs (typically this will be a large number to start with)?
- Which jobs, needing how many people and how much training time, can be filled by local people sent to Japan for relatively short periods?
- Which jobs, needing how many people and how much training time, can be filled by local relatively experienced workers, receiving brief on-the-job training by workers sent from Japan?
- Which jobs, needing how many people and how much time, require long periods of on-the-job-training by workers sent from Japan?
- If it becomes difficult to secure skilled workers, can a job's work content be altered so that half or nonskilled workers can cope?

To answer these questions, a strong grasp is needed of the skill requirements of each job. The keys to success in overseas expansion can be considered as understanding the type of work involved and the differences in domestic and overseas skills levels.

9.2 PROBLEMS AFTER TRANSFER

Problems can occur after completion of technology transfer and the start of production. Among them are workers leaving and the effect of that on a company's skills level. The leakage of secrets to competitors is also a problem. These are the subject of this section.

It has already been described that it is common in overseas expansions for local workers' skill level to start from almost zero. It is a serious setback to a company when such workers leave their job before becoming fully trained. The overseas social system, cultural background, and business culture are all things that can lead to the local workforce stability being different from that in Japan. There are cases overseas in which the technology and skills of local workers have been increased by giving them education and training. The workers have then been sold off to other companies. Or the workers have themselves switched companies.

Apart from this indirect problem, turnover of the workforce directly influences the technical level of a company. To illustrate this consider the relationship between job leaving rate (the fraction of workforce leaving each year) and a company's technical level, assuming that the company is newly set up overseas and also the following:

- The skill level of all employed engineers and technicians is initially zero.
- The workforce number is set at 100, and only workers who leave are replaced.
- The technical levels reached by workers after 1, 2, 3, 4, and 5 years are, on a scale of 0 to 1, 0.25, 0.50, 0.75, 0.9, and 1.0, respectively. (This is approximately the learning curve data for company D in Figure 2.15.)
- The company's technical level is the total of the level of all its engineers and technicians, divided by 100 (i.e., the total number of workers).

Figure 9.1 shows the increase of the company's technical level year by year, from start up in 2008, under different assumptions of job leaving rate (0.1 to 0.4). It is further assumed that the leaving is spread evenly among workers, independently of how long they have been employed. The technical level grows steadily over the first 5 years, independently of job leaving rate, before becoming steady. The 5-year period comes from the assumption that skills grow to their full value over that time. The example indicates that when

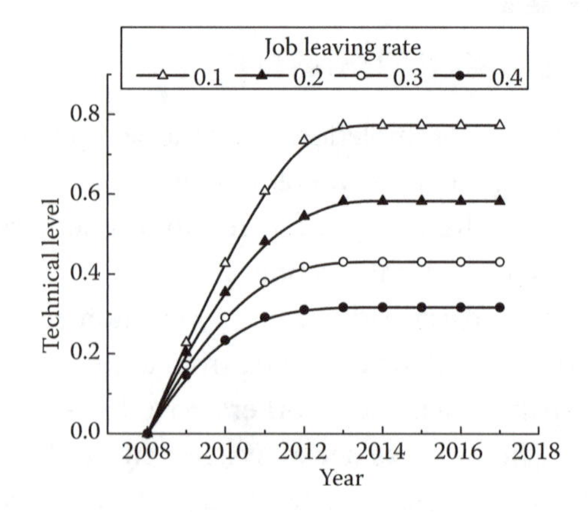

FIGURE 9.1
Increase of the company's technical level with different job leaving rates.

judging the achievements of an expanding company, it is necessary to wait until the years taken for mastering the technology and skill have passed.

Figure 9.1 also shows that the higher the leaving rate, the lower is the technical level that the company can achieve. Table 9.1a records the numbers of workers at each technical level, depending on the job leaving rate, once the steady state has been reached. Table 9.1b shows the job leaving rate of workers at level 0.25 or less, at level 0.5, and at level 0.75 and above, and the average level of workers in the company.

Considering a technical level of 0.25 being that of a beginner and a level above 0.5 being for middle-ranking and expert workers, normal company activity would be difficult with a job leaving rate of 0.4. Then beginners would constantly make up 64% of the workforce. If the job leaving rate drops to 0.3, beginners reduce to 51%. The mid-ranking and expert workers make up the other half. It may be possible for company activities somehow to take place. At a job leaving rate of 0.2, with beginners at 36% (i.e., about one-third) and an average technical level of 0.58, a company will have no major obstacle to its activities. If the job leaving rate becomes 0.1, and the average technical level becomes 0.77, then a company will have a technical level that is similar to or better than those of companies in Japan. These numbers (with a 5-year period) are for the training of skilled engineers and

TABLE 9.1a

Number of Workers at Each Technical Level Once a Steady State Is Reached, Depending on Job Leaving Rate

		Number of Workers				
Technical Level		**0**	**0.25**	**0.5**	**0.75**	**Over 0.75**
Job leaving rate	0.4	40	24	14	9	13
	0.3	30	21	15	10	24
	0.2	20	16	13	10	41
	0.1	10	9	8	7	66

TABLE 9.1b

Number of Workers at Each Technical Level: Summary and Average Data

Job leaving rate	0.4	0.3	0.2	0.1
No. of workers with technical level ≤ 0.25	64	51	36	19
No. of workers with technical level of 0.5	14	15	13	8
No. of workers with technical level ≥ 0.75	22	44	51	73
Average technical level	0.32	0.43	0.58	0.77

technicians. A job leaving rate of 0.2 or less should be a target in fixing the salaries of skilled staff at a higher level than those of basic workers.

As far as the leakage of technical secrets to competitors is concerned, it has previously been considered in Chapter 3, Section 3.3, that this can occur in three ways: leakage by people, leakage of information by itself, and leakage through manufacturing facilities. Any of these ways are possible in overseas expansion. Leakage by people as a result of job switching in particular becomes a problem. Job switching is comparatively rare in Japan because the sense of belonging to a company and the implications of lifelong employment are still strong. This is not the case overseas in a different society where job switching may not be rare.

There are many experiences of Japanese companies that expanded overseas and appointed local staff to positions of responsibility. These staff gained experience in this way and then resigned to start up a rival company. They also took a small number of colleagues with them. The problem is particularly severe in advanced technology cases where the highly trained workers are difficult to replace and the threat from rival manufacture is consequently greater.

Technology licensing also has problems with leakage, although how that happens can be different in detail from leakage by job switching. In technology licensing, the side that receives technology naturally gains complete control of it and has the possibility of developing independently as a rival, even though it breaks the terms of the licensing contract. Historically, many basic technologies, such as steel manufacturing, shipbuilding, and textiles, have expanded in this way. If we accept that technology has a fate whereby outstanding technology expands and weak technology wanes, the question becomes: When is expansion to be restricted and when should it be accepted and promoted?

9.3 OVERSEAS EXPANSION DECISION MAKING USING BLOCK DIAGRAMS

9.3.1 Benefits of Block Diagrams

As shown in Chapter 2, block diagrams can be used to describe and analyze company activities. This applies to overseas expansion activities too. The use of block diagrams assumes the following:

- Company activities are defined by the flow of materials and information.
- Company activities can be divided into blocks.
- Management of materials and information is associated with each block.
- The materials and information change as a result of the management.
- Time is needed for management.
- The time needed for management can be converted into cost.

It is understood that the purpose of managing the materials and information flow is to add value at every stage. Workers' abilities certainly influence the outcomes of management as workers stand in the chain of activities. When a fault occurs, the block diagram can be used to follow the flow of materials and information, to trace the fault to its source. In manufacturing, many of the engineering and technical activities related to materials and information are carried out in parallel. Information sharing is essential to guarantee the quality of manufacturing. Information sharing becomes easy when planning for it is based on the block diagram. Everyday activities are based on the same idea, even if sometimes that is not consciously the case. Particularly, by breaking down every activity into blocks, adding costs for each block becomes possible and the total production cost can easily be accumulated.

The benefit of using a block diagram to help decision making on overseas expansion is that it helps in analyzing which activities have different costs as a result of moving overseas. Figure 9.2 is an example of a manufacturing block diagram. It is the same as Figure 1.4 in Chapter 1, except that

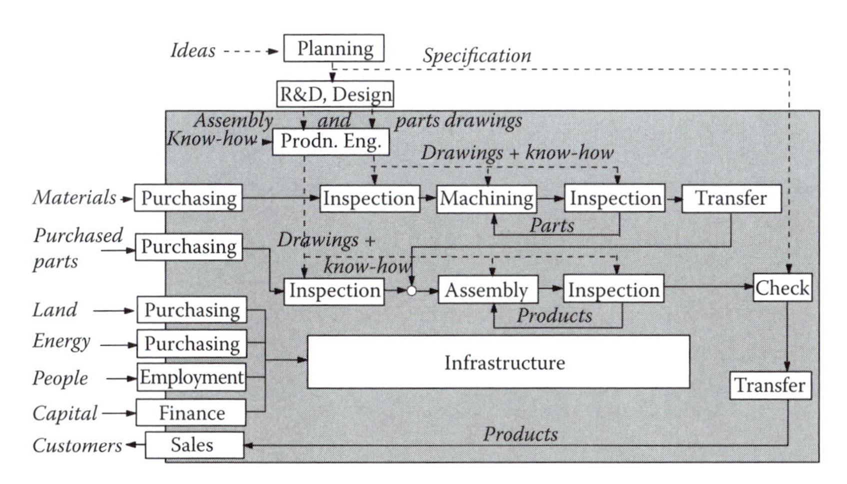

FIGURE 9.2
Information and materials flow in manufacturing.

the background shading has been changed to be only around the blocks concerned with the overseas activity. By following the flow of activities overseas and breaking down the costs by activity, it is possible to capture effectively how beneficial the expansion overseas would be.

9.3.2 A Costing Example, with Quality and Defect Rate Constraints

The thesis of this book is that the quality and defect rate in overseas manufacturing should be the same for a product made overseas as in Japan. To achieve the same quality, it might be necessary to reduce productivity in order to take more care over a process, or to send Japanese workers overseas. It may be more difficult to achieve the same low defect rates. But in any case, there is a cost associated with the targets of equal quality and defect rate. The relative costs of manufacturing domestically or overseas can be estimated, subject to these targets. This provides a rational basis for evaluating whether or where to expand overseas.

Manufacturing cost of a product is calculated as the sum of the price of materials (including purchased parts) and a processing cost. First, how to compute processing cost is described, based on discussions so far.

The manufacturing effectiveness, E_P, of a process depends on its facilities' effectiveness, E_E, the automation rate, R_A, the nonautomation rate, R_{NA}, and the workers' abilities, E_W, as shown in Equation 9.1 (it has the same form as Equation 3.1). E_E, R_A, R_{NA}, and E_W differ from country to country. R_A and R_{NA} add up to 1.0. E_W is obtained from the applicable skills learning curve.

$$E_P = E_E \times (R_A + R_{NA} \times E_W) \tag{9.1}$$

For Japan, EE and EW are set equal to 1.0 so $E_P = 1.0$ too. Then E_P for the target country is a relative effectiveness. The target country's relative processing cost, C_P, is then determined from its relative time cost, C_T, by means of Equation 9.2.

$$C_P = C_T/E_P \tag{9.2}$$

The total processing cost for a product is calculated by summing the costs of the processes involved. Adding the costs of purchased parts

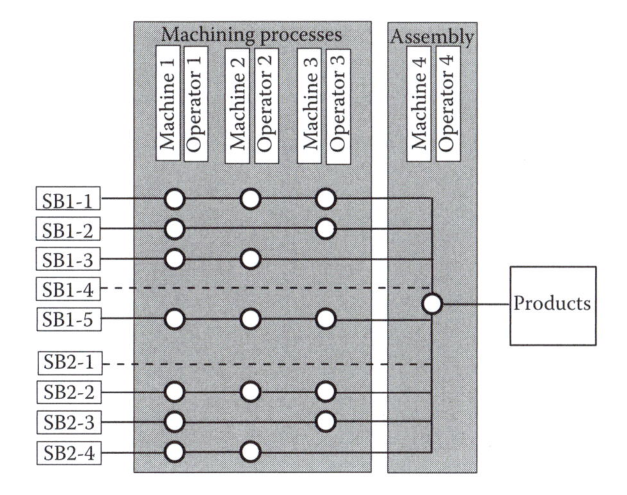

FIGURE 9.3
An operation flowchart example.

and materials to the processing cost gives the manufacturing cost of the product.

A case for overseas expansion is now considered, based on these ideas. It takes a simple fictitious manufactured product as an example. The target product and outline of its manufacturing stages are shown in Figure 9.3. The product is made up of a total of nine parts, SB1-1 to SB1-5 and SB2-1 to SB2-4. Some of the manufacturing stages are machining and some are assembly. Parts SB1-4 and SB2-1 are purchased and go to final assembly without any machining. The other parts need machining before assembly. Machining involves three machine tools. Some parts do not need all the machine tools.

First, the domestic evaluation table is drawn up (Table 9.2). Unit costs of materials, processing cost rates (the sum of workers' cost rates and facilities' cost rates, including depreciation, fuel and lighting costs, and other overheads), and time allocated per part need to be known for this. In addition, the processing cost of each part and processing stage is calculated from the stage cost rates and allocated times. The last row of the table is the manufacturing cost, the sum of processing, and materials/purchasing costs. All prices are converted to U.S. dollars.

Next, this product is assumed to be manufactured in country A overseas. The major features when expanding to country A are assumed as follows:

TABLE 9.2

An Example of Evaluation When Manufacturing in Japan

				Amount of Time (h)				
	Parts	**Buy-In**	**Price (USD)**	**Machine 1 (80 USD/h)**	**Machine 2 (20 USD/h)**	**Machine 3 (40 USD/h)**	**Assembly (120 USD/h)**	**Time Cost (USD)**
SB1	SB1-1	No	100	3.0	4.5	1.0	0.0	370
	SB1-2	No	12	2.5	0.0	2.5	0.0	300
	SB1-3	No	30	0.8	2.3	0.0	0.0	110
	SB1-4	Yes	500	0.0	0.0	0.0	0.0	0.0
	SB1-5	No	40	1.5	3.4	3.5	0.0	328
	Subtotal	—	**682**	**7.8**	**10.2**	**7.0**	**0.0**	**1108**
SB2	SB2-1	Yes	500	0.0	0.0	0.0	0.0	0.0
	SB2-2	No	50	6.0	3.0	4.5	0.0	720
	SB2-3	No	80	4.5	0.0	5.0	0.0	560
	SB2-4	No	30	2.5	4.5	0.0	0.0	290
	Subtotal	—	**660**	**13.0**	**7.5**	**9.5**	**0.0**	**1570**
Assembly	—	—	—	—	—	—	16.0	**1920**
Total time (h)	—	—	—	20.8	17.7	16.5	16.0	—
Cost (USD)	—	—	**1342**	1664	354	660	1920	**4598**
Grand total (USD)					**5940**			

TABLE 9.3

Parameters for Manufacturing in Country A

	Machine 1	Machine 2	Machine 3	Assembly
E_E	0.90	0.70	0.95	0.80
R_A	0.80	0.60	0.70	0.40
R_{NA}	0.20	0.40	0.30	0.60
E_W	0.80	0.70	1.00	0.70
E_P	0.86	0.62	0.95	0.66
C_T	0.85	0.65	1.20	0.60

- In machining stage 3, skills are needed and securing quality is difficult. Hence, workers are sent in from Japan.
- SB1-5 is manufactured in Japan.
- The parameters relating to manufacturing effectiveness and time costs in country A are as shown in Table 9.3.
- The prices of materials and purchased parts in country A are different from those purchased in Japan.

Based on these criteria, the evaluation table for expanding to country A is shown in Table 9.4. From this result we can predict that it is not cost-effective to expand overseas to country A.

Now assume overseas expansion to country B. Compared to country A, the features of overseas expansion to country B are:

- The quality of processing stage 3 can be secured. However, work efficiency is low.
- Like country A, SB1-5 uses processed parts from Japan.
- The parameters relating to manufacturing effectiveness and time costs in country B are as shown in Table 9.5. The time costs and workers' abilities are higher in country B than in country A.
- The price of materials and purchased parts in country B are different from those purchased in Japan and country A.

Based on these criteria, the evaluation table in the case of overseas expansion to country B is shown in Table 9.6 (page 168). If overseas expansion to country B were to be carried out, it would be more beneficial in terms of costs than manufacturing in Japan. From these considerations, country B is a better option than country A to be targeted for the

TABLE 9.4

An Example of Evaluation When Manufacturing in Country A

				Amount of Time (h)				
	Parts	**Buy-In**	**Price (USD)**	**Machine 1 (68 USD/h)**	**Machine 2 (13 USD/h)**	**Machine 3 (48 USD/h)**	**Assembly (72 USD/h)**	**Time Cost (USD)**
SB1	SB1-1	No	105	3.47	7.31	1.05	0.0	381.4
	SB1-2	No	11	2.89	0.0	2.63	0.0	322.8
	SB1-3	No	28	0.93	3.73	0.0	0.0	111.8
	SB1-4	Yes	490	0.0	0.0	0.0	0.0	0.0
	SB1-5	Yes	370	0.0	0.0	0.0	0.0	0.0
	Subtotal	—	**1004**	**7.29**	**11.04**	**3.68**	**0.0**	**815.9**
SB2	SB2-1	Yes	550	0.0	0.0	0.0	0.0	0.0
	SB2-2	No	48	6.94	4.87	4.74	0.0	762.7
	SB2-3	No	88	5.21	0.0	5.26	0.0	606.8
	SB2-4	No	27	2.89	7.31	0.0	0.0	291.6
	Subtotal	—	**713**	**15.04**	**12.18**	**10.00**	**0.0**	**1661.1**
Assembly time	—	—	—	—	—	—	24.39	**1756.1**
Total time (h)	—	—	—	22.34	23.21	13.68	24.39	—
Cost (USD)	—	—	**1717**					4233.1
Grand total (USD)				**5950.1**				

TABLE 9.5

Parameters for Manufacturing in Country B

	Machine 1	Machine 2	Machine 3	Assembly
E_E	0.95	0.80	0.95	0.85
R_A	0.80	0.60	0.70	0.40
R_{NA}	0.20	0.40	0.30	0.60
E_W	0.85	0.75	0.50	0.80
E_P	0.92	0.72	0.81	0.75
C_T	0.90	0.70	0.75	0.65

manufacture of the example product. Using the methods described here, by considering the technical level (men and facilities) and economic situation of each country, advantages of overseas expansion can be evaluated quantitatively.

DISCUSSION QUESTIONS

1. Overseas technology transfer always involves the risk of leaks of technical intelligence. How may the risk be kept to a minimum?
2. If you have a new advanced technology but it is impossible to keep a monopoly on it, in what circumstances should you consider transferring it?
3. Job leaving rate is a large issue in overseas production. What is the maximum allowable job leaving rate when a worker needs 2 years to become fully skilled?

TABLE 9.6

An Example of Evaluation When Manufacturing in Country B

				Amount of Time (h)				
	Parts	**Buy-In**	**Price (USD)**	**Machine 1 (72 USD/h)**	**Machine 2 (14 USD/h)**	**Machine 3 (30 USD/h)**	**Assembly (78 USD/h)**	**Time Cost (USD)**
SB1	SB1-1	No	99	3.26	6.25	1.24	0.0	359.4
	SB1-2	No	13	2.71	0.0	3.10	0.0	288.1
	SB1-3	No	28	0.87	3.19	0.0	0.0	107.3
	SB1-4	Yes	495	0.0	0.0	0.0	0.0	0.0
	SB1-5	Yes	370	0.0	0.0	0.0	0.0	0.0
	Subtotal	—	**1005**	**6.84**	**9.44**	**4.34**	**0.0**	**754.8**
SB2	SB2-1	Yes	520	0.0	0.0	0.0	0.0	0.0
	SB2-2	No	52	6.51	4.17	5.57	0.0	694.2
	SB2-3	No	88	4.88	0.0	6.19	0.0	537.1
	SB2-4	No	35	2.71	6.25	0.0	0.0	282.6
	Subtotal	—	**695**	**14.10**	**10.42**	**11.76**	**0.0**	**1513.9**
Assembly time	—	—	—	—	—	—	21.39	**1668.4**
Total time (h)	—	—	—	20.94	19.86	16.10	21.39	—
Cost (USD)	—	—	**1700**					**3937.1**
Grand total (USD)					**5637.1**			

10

Example of Shipbuilding Industry in Overseas Technology Transfer

This chapter is about problems of technology transfer in industries in which tacit knowledge (know-how) plays a large role. The shipbuilding industry is taken as an example. In the next introductory paragraphs differences in the knowledge requirements of shipbuilding and mass production industries are considered. These then lead to a discussion of the factors affecting the global competitiveness of technology transfers in shipbuilding. These are wider than just technical factors. After that, examples of success and failure of technology transfer in the shipbuilding industry are described. These are then used to extract what are the important points concerning technology transfer and knowledge requirements.

Generally, manufacturing of products is carried out by using equipment such as factories, facilities, and tools, as well as by human work. One direction of equipment development is to mechanize tasks that used to be carried out by people. Production today makes full use of such developments. Development of mass production breaks down manufacture into tasks and arranges these along conveyor lines. This makes mechanization and automation easier. It simplifies the human function and reduces dependency on human skills. In an automated factory, the equipment is in charge of most of the processes. The roles of humans are to check the equipment or administer the production work. Automated manufacturing processes are very different from previous ones because of their very different dependences on people and equipment. These developments result in a relatively greater dependence now on equipment than on high-level human skills.

In technology transfer the division of functions between equipment and people is rebuilt in a different environment. Particularly, the question how to combine human labor and equipment under the social and economic environment of another country has to be answered. In today's globalized economic environment, there is a trend to equal equipment costs in different countries. However, labor costs show remarkable differences. They depend on the economic environment of the country and are not transferable like equipment. The differences in human labor have become the fundamental driving force in the present transfer environment. The equipment vs. human labor needs of manufacturing a product have a big influence on the costs and benefits of a technology transfer and how a transfer is carried out.

The shipbuilding industry is typically an individual order industry. It is necessary to design and build a complicated system in a short period of time. This requirement has a big influence on design and production. In the case of mass-produced goods like cars, etc., it is common that detailed manuals about working methods, etc., are prepared. In addition, introducing automatic production tools tends to reduce costs and manufacturing time. However, for individual orders, even if operation manuals and automatic production tools were to be prepared, they could only be used on one product. They would be ineffective in many cases. Furthermore, the manual tasks and hence automation become much more difficult as a product becomes more complicated. The most effective method of production in this kind of case is to secure excellent engineers and workers who can work appropriately and efficiently by themselves, without being given detailed instructions.

This has consequent problems for technology transfer. In the case of mass-produced products, the first steps are the transferring overseas of the manufacturing equipment and the training and positioning of workers who can operate according to a manual's instructions. Judging from experience, technology transfer in such cases is comparatively easy. However, in the case of shipbuilding, technology transfer that does not involve equipment and manuals is important as well. As it is a labor-intensive type of industry, transfer is a natural move from the point of view of reducing labor costs. However, overseas transfers of shipbuilding have not been often attempted because the transfer of individual experience and know-how is difficult. Even when attempted, it is often not carried out successfully.

The term *tacit knowledge* was used at the start of this chapter. It has been used in previous chapters but can be explained again. The human role in manufacturing is to intervene to process and transform products effectively. Knowledge is required to decide and judge how best to do that. Although there is a tendency to develop knowledge support systems, humans themselves still play a vital role. The knowledge can generally be classified into two types:

- Explicit knowledge: Formal knowledge that is easy to be transferred through, for example, standard instructions or logical language.
- Tacit knowledge: Knowledge that is based on personal experience and which is difficult to communicate to others.

As has been written in the previous paragraphs, technology transfer of mass-produced goods starts with the transfer of explicit knowledge, while shipbuilding industry technology transfer requires more transfer of tacit knowledge. Here, problems that stem from that are considered.

10.1 GENERAL SURVEY OF SHIPBUILDING TRANSFERS AND SELECTION OF SUCCESSFUL AND UNSUCCESSFUL CASES

It is generally true of overseas expansion of manufacturing industry that the relative competitiveness between domestic and overseas manufacturing depends in part on how the manufacturing is managed. Attention must be paid to management issues. There are cases where transfers have failed because technology and methods existing in Japan have been transferred without modification. Here these and other items that need to be considered in technology transfer of shipbuilding are considered. The material is based on the results of two comparative surveys of the shipbuilding industries in Indonesia, Malaysia, Singapore, and Japan. The surveys were led by Hiroshima University with the support of the Japan Society for the Promotion of Science.

10.1.1 Comparison Measures

Shipbuilding industry comparisons were made in terms of a shipyard's business and financial state, manufacturing capabilities and facilities, production planning and control, human resources and organization, and support systems. These five main headings were broken down into 17 subheadings and then into a further 44 items in all, as listed in more detail in Table 10.1.

TABLE 10.1

Comparison Areas

Category (Level 1)	Category (Level 2)	Item
Shipyard's business and financial state	Principal measurement scale	Profit rate Quality level Delivery estimation Delivery delay
	Market	Future demand estimation Marketing activity
	Finance	Fund-raising forecast Fund-raising activity
	Trust	Customer (owner) trust Supplier trust
	Competitiveness	Competitiveness level
Manufacturing capabilities and facilities	Capacity	Facility capacity level Manpower capacity level
	Process and technology	Production technology level Production lead time CAD/CAM
	Product scope and new products	New product development level Production flexibility
	Facilities and plant	Condition of facilities Facility layout Plant space
Production planning and control	Production planning	Production planning and scheduling reliability Frequency of changes in planning
	Production control	Shop floor control Workforce allocation Material handling
Human resources and organization	Human resources	Knowledge of the staff Skill of the workers Motivation level Turnover rate
	Organization	Autonomy delegation Training and education level Team work strength

(continued)

TABLE 10.1 (CONTINUED)

Comparison Areas

Category (Level 1)	Category (Level 2)	Item
Support systems	Work measurement	Workers' productivity
		Wages
		5S[a]
		Safety recognition
	Supplier relations	Parts delivery date/materials procurement
		Lead time of parts/materials procurement
	Information systems	Computer application
		Database management
		Information technology (IT) application
	Improvement activities and motivation	Level of improvement activities
		Improvement motivation

[a] 5S Select (what is needed), Set (in order), Shine (keep clean), Standardize, Sustain (enthusiasm).

For each shipyard surveyed, five of its engineers or managers and five Japanese university staff evaluated its performance in terms of Table 10.1's items. The 5 countries, 6 regions, and 14 shipyards that were the subject of the survey are described in more detail in the following, including the nature of the shipyards' businesses and environment in each country.

- Indonesia (Java region): Two shipyards were selected from Indonesia's Java Island. These shipyards are recognized as the main shipyards in Indonesia and handle various types of ships.
- Indonesia (Batam region): Batam is a region opposite Singapore. It is well placed to buy ship equipment from suppliers in Singapore. In addition, it specializes in ship repair and in new building of both simple and specialized types of ships. Three shipyards from the Batam region were selected.
- Malaysia: One shipyard from Malaysia was selected. This shipyard is one of the main shipyards in Malaysia. The main part of its business is repair work.
- Philippines: As one of the successful examples of technology transfer, one shipyard is selected from the Philippines. The parent company in Japan takes charge of activities that require highly developed

technology, such as design, production planning, procurement, and orders.

- Singapore: Most of the shipyards in Singapore have repair work as their main business. However, construction of new ships is also carried out because of the close connection with Singapore's shipping industry. Three shipyards were selected from Singapore.
- Japan: Considering the size of the shipyards of the South East Asian countries in the survey, four middle-sized shipyards from Japan were selected. Compared to the other countries' shipyards, the number of workers is very small in the Japanese yards, and the technology level is high.

10.1.2 Survey Results

Table 10.2 shows the survey results at the 17 subdivision levels. For each region and subdivision there are three numbers. The upper one is the internal evaluation by engineers and workers. The lower one is the external evaluation by the Japanese university staff. The middle one is the difference between the two. The evaluations were made on Likert-type scales, from 1 (worst) to 5 (best). From the table, the following points can be observed:

- According to the internal evaluations by company engineers and workers, the performance of the Batam and Philippine shipyards is slightly better than that of the others, although the differences are insignificant.
- According to the external evaluations by Japanese university staff there are significant differences between countries. The overall ranking/order based on these results is Philippines >> Japan = Singapore > Malaysia > Batam >> Java.
- From the standpoint just of competitiveness, the ranking can be summarized as Philippines >> Japan = Singapore = Batam > Malaysia >> Java.
- There are differences between countries as far as comparisons between internal and external evaluations are concerned. The differences between evaluations from companies' engineers and university staffs for Japan and Singapore are comparatively small. However, the Philippines engineers tend to appraise themselves lower than the study group's members, while Batam companies' engineers tend to appraise themselves higher. This tendency is even greater in Java.

10.1.3 Selections of Successful and Unsuccessful Cases

Based on the foregoing discussions, it can be judged that promotions of shipbuilding in the Philippines have been successful, while promotions of shipbuilding in Indonesia might not have been. Moreover, all the shipyards concerned have received various supports from Japan. Therefore, in this chapter these shipyards are selected as successful and unsuccessful cases of technology transfer in shipbuilding.

Within Indonesia there is an interesting difference, from Table 10.2, between Java and Batam. According to external evaluation, Batam shipyards are relatively highly regarded as far as their competitiveness is concerned, although most other evaluations are low. The following are characteristics of the Batam shipyards that might strongly affect their competitiveness:

- They specialize in ship repair and in new building of both simple and specialized types of ship (construction strategy).
- They can source ship equipment from suppliers in Singapore and their supply chain works well (supply chain).
- In Table 10.2, the evaluation result for organization is comparatively high (motivation and administration).

Based on these points, two real examples of technology transfer are introduced in Sections 10.2 and 10.3, paying attention to the following points:

- Problems of tacit knowledge: How should the tacit knowledge of skilled workers be transferred?
- Problems of construction strategy: How should low levels of technical abilities be supplemented, and what construction strategies should be set in order to accumulate construction knowledge and know-how (including outsourcing of design and other things)?
- Problems of the supply chain: How should an appropriate supply chain be secured?
- Problems of motivation and administration: How should motivation in local workers and engineers be instilled? How should management methods for workers in the local area be developed?

TABLE 10.2

Survey Results (internal and external regional assessments differing by >0.5 are shaded light gray (internal > external) or darker gray (external < internal))

		Shipyard's Business and Financial State					Manufacturing Capabilities and Facilities				Production Planning and Control		Human Resources and Organization		Support Systems				
		Principal Measurement Scale	Market	Finance	Trust	Competitiveness	Capacity	Process and Technology	Production Scope and New Products	Facilities and Plant	Production Planning	Production Control	Human Resources	Organization	Work Measurement	Supplier Relations	Information Systems	Improvement Activities, Motivation	Average
Java	Internal	2.78	4.15	3.70	3.25	3.70	3.15	3.07	3.10	2.73	2.85	2.90	3.25	2.57	3.34	2.90	2.87	2.90	3.13
	Difference	1.44	2.61	2.33	2.08	2.20	1.44	1.40	1.93	0.71	1.60	1.68	1.81	0.79	1.86	1.73	0.59	1.57	1.61
	External	1.33	1.54	1.38	1.17	1.50	1.71	1.67	1.17	2.03	1.25	1.22	1.44	1.78	1.48	1.17	2.28	1.33	1.52
Batam	Internal	3.30	3.83	3.85	3.80	3.72	3.45	3.08	3.47	3.34	3.31	3.33	3.09	3.14	3.21	3.40	3.10	3.20	3.33
	Difference	0.59	0.85	0.93	0.73	0.36	0.48	0.67	1.00	0.37	0.67	0.69	0.44	0.15	0.48	0.41	0.72	0.88	0.59
	External	2.71	2.98	2.92	3.07	3.36	2.97	2.41	2.47	2.96	2.64	2.64	2.65	2.99	2.73	2.99	2.37	2.32	2.74
Malaysia	Internal	2.88	2.88	2.75	3.63	3.25	2.75	2.67	2.17	3.25	2.25	2.92	2.56	3.42	2.69	3.50	2.50	2.38	2.83
	Difference	0.03	−0.23	−0.88	0.63	0.65	−0.45	−0.23	−1.03	0.18	−0.65	−0.25	−0.66	0.53	−0.06	0.63	0.37	−0.43	−0.12
	External	2.85	3.10	3.63	3.00	2.60	3.20	2.90	3.20	3.07	2.90	3.17	3.23	2.88	2.75	2.88	2.13	2.80	2.95

Philippines	Internal	4.11	3.98	3.43	4.01	4.50	3.59	3.32	3.50	3.09	3.49	3.37	3.37	3.24	3.51	3.24	3.17	2.95	3.49
	Difference	−0.31	0.65	−0.24	−0.16	0.17	−0.91	−0.57	1.00	−0.80	−0.85	−0.74	−0.88	−0.65	−0.15	−0.43	0.06	−0.22	−0.35
	External	4.42	3.33	3.67	4.17	4.33	4.50	3.89	2.50	3.89	4.33	4.11	4.25	3.89	3.67	3.67	3.11	3.17	3.84
Singapore	Internal	3.20	3.28	3.16	3.79	3.65	3.35	2.91	2.93	3.25	2.68	2.87	3.13	3.08	3.35	3.46	2.84	2.89	3.14
	Difference	0.08	−0.15	−0.18	0.23	0.38	0.12	0.04	0.21	0.43	−0.28	−0.09	0.11	−0.07	0.33	−0.03	0.48	−0.11	0.10
	External	3.11	3.43	3.34	3.57	3.27	3.23	2.87	2.72	2.82	2.96	2.97	3.02	3.14	3.01	3.48	2.37	3.00	3.04
Japan	Internal	3.33	2.83	3.14	3.82	3.11	2.90	3.05	3.15	3.08	2.87	2.95	3.01	2.49	2.73	3.53	2.75	2.57	3.00
	Difference	−0.21	0.20	0.27	0.07	−0.39	−0.48	0.13	0.15	−0.17	−0.30	−0.22	−0.29	−0.34	−0.41	0.03	0.80	−0.35	−0.10
	External	3.54	2.63	2.88	3.75	3.50	3.38	2.92	3.00	3.25	3.17	3.17	3.29	2.83	3.15	3.50	1.94	2.92	3.09
Average	Internal	3.26	3.49	3.34	3.72	3.66	3.20	3.01	3.05	3.12	2.91	3.06	3.07	2.99	3.14	3.34	2.87	2.81	3.15
	Difference	0.27	0.65	0.37	0.60	0.56	0.03	0.24	0.54	0.12	0.03	0.18	0.09	0.07	0.34	0.39	0.50	0.22	0.29
	External	2.99	2.84	2.97	3.12	3.09	3.16	2.77	2.51	3.00	2.87	2.88	2.98	2.92	2.80	2.95	2.37	2.59	2.86

10.2 CASE STUDY 1: TSUNEISHI HEAVY INDUSTRIES

10.2.1 Background to Overseas Expansion

Here the transfer of technology by Tsuneishi Shipbuilding of Japan to Tsuneishi Heavy Industries (Cebu), Inc. (THI) in the Philippines is introduced as an example. Tsuneishi Shipbuilding carried out overseas expansion for two reasons.

The first was the sudden appreciation of the yen starting from the late 1980s. Contracts in shipbuilding are mostly dollar based and income is in dollars. But construction costs in Japan are in yen. Even a one-yen appreciation causes profit to fall by hundreds of millions of yen and greatly affects a company's management. Turning construction costs into dollars is the most effective measure against appreciation of the yen. The exchange rate risks disappear since the income and expenditure are both in dollars. As long as shipbuilding was carried out in domestic Japan, it was necessary to pay procurement costs, such as of steel materials, and employees' salaries in yen. Although it could have been possible to supply, for example, materials from overseas, it could not be said to be the best policy considering the high costs of transportation.

The second reason was the rise of Korea and China in the shipbuilding industry. Japan became number one shipbuilder in the world in 1956, in terms of number of ships built. It held that position for many years. However, in recent years that position passed to Korea, and it is obvious that in the near future China will catch up. In order to compete with shipbuilding industries in Korea and China, expansion to a country with cheap labor costs was essential.

These are the reasons why Tsuneishi Shipbuilding expanded overseas.

10.2.2 Selection of the Place

The countries researched for overseas expansion included China, Indonesia, Malaysia, and Vietnam. Finally, in September 1994, Tsuneishi Shipbuilding set up THI in the Cebu Islands of the Philippines based on the reasons below:

- English is the official language and communication could be made easily.

- There is cheap and abundant labor, and the education level is also high.
- In the Philippines, diligence is a nationality trait, particularly in the Visaya region of Cebu Island.
- It was thought that the people would be comfortable with Japanese customs and culture.
- Geographically, Cebu Island is situated in the middle of South East Asia with easy access to Japan, Hong Kong, and Singapore.
- There was strong support for expansion from the local state government.
- There was a cooperation system in place for meeting local partners.

10.2.3 Selection of Local Partners

THI is a joint venture company. Tsuneishi Shipbuilding has an 80% stake. The local Spanish combine, Aboitiz Group, has a 20% stake. Aboitiz Group is a conglomerate enterprise that represents the Republic of the Philippines.

When establishing a company in a developing country like the Philippines, there is the cost benefit of cheap labor, but there is also a big risk. The nature of the locality with its different language, culture, and customs may result in failure of the venture even if the Japanese use Japanese management methods. This risk was reduced in THI by Tsuneishi Group taking charge of the management of technical aspects of shipbuilding, while the Aboitiz Group took charge of the workers. Through this division of responsibilities, there has been no major company management problem to date.

10.2.4 Technology Transfer in THI

The proposition from Tsuneishi Shipbuilding to THI was "to launch a 23,000-ton-type bulk carrier in 2 years from starting construction of the shipyard." At that time, the construction site of the shipyard had just been selected and the shipyard itself did not exist. Therefore, how to build the ships had to be considered at the same time as the shipyard was being constructed. All of the following had to be developed at the same time:

- Infrastructure such as roads, waterworks, and high-tension power transmission.
- Physical aspects such as factory building, cranes, and the introduction of various types of machinery.

- Human relations developments such as creating friendly relations with local businesses, recruitment strategy, and personnel training.

The most important investment in the future was recruitment of the first 15 talented employees who would carry the future of THI. These were each paired with a Japanese construction supervisor. They participated in each of the constructions that they would be in charge of in the future. During this construction period, trust toward the Japanese was built. It was also a success in the sense that the workers felt responsible as they built the facilities themselves.

In parallel with the shipyard's construction, shipbuilding training also advanced, considering the needs of further developments in the future.

First, the 15 initial workers were invited to Japan and were given a half year's training. Once the shipyard was built, they would further learn shipbuilding methods on-site while working with Japanese engineers. Also, a relationship of trust with the Japanese staff was built by working together.

Second, when commissioning the shipyard, approximately 100 engineers and workers were sent from Japan in order to transfer advanced technology quickly. Although it is easy to communicate in the Philippines, as English is used as a common language, at that time about 90% of the instructors could not speak English. Further, how to teach on-site skills was not established either. An instructor would first demonstrate how to make something himself and let it be copied.

Third, the policy was for the workers to understand shipbuilding technology and to learn by trial and error until they could work on their own. It was decided to specialize in the building of handy-size bulk carriers. When the operation of the shipyard started, the building of 23,000-ton bulk carriers (23BC) took about 4 months. Now, 52,000-ton-type bulk carriers (52BC) can be launched in a little less than 40 days. This is a result of the continuous production techniques, specializing in a single type of ship, that the workers developed.

Year-on-year increases in productive capacity, sales, and the number of workers are shown in Table 10.3. The table also shows that while the number of sales and construction increased favorably, the number of Japanese workers decreased. Hence, it can be concluded that this example is one of successful technology transfer.

TABLE 10.3

Year-on-Year Increase of Activity at THI

Year	New build: 23–52 ktonne Bulk Carriers	Repairs	Sales (Mpeso)	Employees	Japanese Advisors		
					Managers	Skilled Workers	Total
1997	23BC × 3	42	1678	1789	30	68	98
1998	23BC × 4	62	2461	2119	38	22	60
1999	23BC × 2, 45BC × 2	92	3176	2055	27	2	29
2000	28BC × 2, 45BC × 3	93	4207	3031	25	2	27
2001	45BC × 2, 52BC × 5	111	6085	3963	24	1	25
2002	52BC × 7	134	5953	3827	19	1	20
2003	52BC × 7	103	6334	3671	21	0	21
2004	52BC × 7	77	7504	4233	20	0	20

10.3 CASE STUDY 2: TECHNICAL COOPERATION IN SHIPBUILDING TO INDONESIA

Shipping and shipbuilding business are very important industries to Indonesia. Indonesia is the biggest maritime country in the world. Since the founding of the country, development of these industries has been pursued as a high national policy. Some important conclusions on technology transfer within large-scale industries like shipbuilding can be made by following developments in Indonesia.

10.3.1 Outline of Indonesia's Shipbuilding Industry

The main shipyards in Indonesia are four national shipyards and five private shipyards. Although more than 200 registered shipyards and 60 companies are members of the Indonesia Offshore Industry and Shipping Association (IPERINDO), more than half of them are small shipyards mainly for repair work. Nevertheless, when considering the social infrastructure of the archipelagic country, which has more than 17,000 islands and a population of more than 200 million people, this number of facilities is extremely small for the maintenance of the 2100 ships of Indonesian nationality totalling 7 million deadweight tonnage (DWT).

Most domestic shipping companies operate with used ships rather than with new build. Between the economic crises of 1997 and 2003 there was no demand for private shipbuilding companies. However, after the international revision of the Load Line Rules in 2003, there was a new shipbuilding boom. Indonesian yards received orders for new ships.

As far as repair is concerned, despite there being huge domestic demands, there is a low productivity, with repairs typically taking 1 month. This leads to losses and to most of the leading shipping companies of Indonesia having their own repair facilities.

10.3.2 Development of Indonesian Shipbuilding Industry

The development of Indonesian shipbuilding was initially successful. However, there were subsequent failures in promoting new shipyards. This, as well as a slump in customers for marine transportation, has led to a current inactivity.

10.3.2.1 An Initial Success Story (the Origin of the Indonesian Shipbuilding Industry)

The story of Indonesian new shipbuilding began in 1982 when Pertamina placed orders with domestic shipyards for five 3500DWT-type tankers. Building these ships, as well as most other new ships through the 1980s, was supported by design and construction instructions from Japanese shipyards. Procurement of parts was by the Japanese yards and also by trading companies, as well as through package deals. Although the supporting parties of Japan and others were responsible for the design, material supplies, and management of the construction process, ships were successfully supplied in the sense that the local site was in charge of the yards and supply of workers.

10.3.2.2 The Caraka Jaya, Mina Jaya, and Other Projects

The Indonesian government started the Caraka Jaya project in 1988. Its purposes were to raise the abilities of shipyards and to develop Indonesian marine business, by means of supporting the building of new ships suitable for marine business between islands. The project can be divided into three periods, the first from 1988 to 1991, the second from 1991 to 1994, and the third from 1994 until now. Export credit from Germany and Japan was used as funding, with PT.PAL in charge of the design of the ships. In each phase, series construction of the same types of ship was planned in order to learn about the construction technologies.

In the first period, a 3000DWT-type semicontainer ship and a general cargo ship were built at PT.PAL's shipyard, and three general cargo ships of the same type were built in three shipyards in Jakarta. These three shipyards would later become merged as DKB.

In the second period, a total of 27 ships was built: 2 of 3000DWT-type and 11 of 3650DWT-type semicontainer ships built at PT.PAL, 2 of 3650DWT-type semicontainer ships built at DPS (Dok Surabaya), and 12 of 3650DWT-type general cargo ships built at DKB and two private shipyards. Although these ships were completed, there were large delays in construction. But with the help of the public corporation specialized fund management company PT.PANN a profit was maintained.

In the third period, construction of twenty-four 4180DWT-type semicontainer ships was started in nine shipyards, including the main private shipyards. The first nine of the ships were completed, although construction took longer than expected as the series construction methods did not progress well. However, as a result of financial difficulties brought on by the 1997 economic crisis, as well as bad management by PT.PANN, construction of the 15 remaining ships was halted. The imported building materials for 15 ships, such as main engines, cranes, etc., were left unused at the shipyards.

The so-called Mina Jaya project was another shipbuilding project with government guidance. Thirty-one 300GT-type fishing boats were to be built as a batch at the IKI shipyard in Makassar, funded by a loan package from Spain. The Spanish side quickly sent most of the parts for assembly, including the hull in prefabricated blocks and engine-related parts. It was beyond the ability of the IKI shipyard to deal with these. As a result of the delays in assembling them and at the same time a decrease in demand for them, most of the ships were left unsold. Nine of the completed ships were left moored to the dock, with huge blocks of unfinished hulls only partly assembled occupying most of the shipyard. These were not reused.

The Indonesian Shipbuilding Industry has also failed in construction of export ships. At the beginning of the 1990s, four of the national shipyards in Jakarta port were combined and DKB was inaugurated. Although at first it was a steady shipyard that focused on repair work, it gradually increased the amount of its new shipbuilding. After the integration of the yards, it concentrated on new, larger-type shipbuilding and set about ship exporting. In the end, this backfired and brought about large financial losses due to penalties for construction delays and low quality.

10.3.3 Japanese Assistance to Indonesian Shipbuilding Industry

Japan's assistance to the Indonesian shipbuilding industry began from early times, by accepting foreign students as part of postwar compensation. To this day these students play an important part in universities, IPERINDO, private shipbuilding, and related industries. Over the long term, they have a significant influence and importance in the area of personnel training cooperation. In addition, in 1979, the Overseas Economic Cooperation Fund of Japan (OECF JAPAN) financially supported the facilities expansion plan (7000DWT-type dock, steel processing plant, and related facilities) of PT.Pelita Bahari. This was one of the four shipyards that later integrated into DKB, the current national shipyard. From 1979 to 2000 the Japan International Cooperation Agency (JICA) sent more than 50 private shipbuilding specialists to four of the national shipyards. After that, shipbuilding specialists were sent as senior volunteers.

10.3.4 Problems of Indonesian Shipbuilding Development

10.3.4.1 Problems of National Projects

Indonesia's Caraka Jaya project is said to be the same type of shipbuilding promotion that Japan used for the development of its shipbuilding industry. After the war, the shipbuilding industry of Japan was a marginal producer compared to Europe and America. The promotion of shipbuilding in Japan succeeded by introducing new facilities and technologies to meet the new demands for large-sized ships. In addition to these strategies, Japanese shipyards already had a high-technology potential from their prewar expertise. Moreover, after the war there was an almost guaranteed demand for the ships, although that could have reduced with unexpected economic changes. Even when demand reduced, a minimum demand for ships was guaranteed by the government to promote domestic logistics. Since the shipbuilding industry in Japan quickly acquired the ability to compete on equal terms in a fierce international market, these policies can be said to have succeeded.

However, in the 1990s, when Indonesia started the promotion of its shipbuilding industry, it was exposed to severe cost competition because of overcapacity in the industry. In addition, the international economy was shifting to being driven by the advancing industrialization in Asia. In these conditions, it was required not only to build a ship, but to build it to globally competitive quality and cost standards.

Because of the rapid progress of containerization, the ships being built in the Caraka Jaya project were already missing the trend of the shipping market. The introduction of technology is only one of the conditions of industrial development. To be successful, a policy of promoting industries must also understand the economic environment of the time.

10.3.4.2 Problems of Alienation from the Needs of the Shipping Industry

For Indonesia, development of shipping trade is important so that people and industry can carry out their daily lives. However, new shipbuilding, in particular of large-sized ships, does not meet the needs of Indonesia's shipping.

Interest rates in developing countries are high, and Indonesia is no exception to that. Because of the interest rates, most of the private shipping companies of Indonesia purchase small used ships from the international market and maintain and use them appropriately. The use of such used ships is common in a developing country. Although the ships required for operation are purchased from the international market, the repair work is done locally to cut down the out-of-service time. Hence, development of the repair industry is important as well.

From the above, from the point of view that the shipbuilding industry should be a support industry for the shipping industry, it is considered that development with emphasis on repair work is more important than building new ships. Also, as far as building new ships is concerned, construction of coastal vessels should be considered because it is difficult to purchase these on the international market.

10.3.4.3 Management Problems

The failures of the Caraka Jaya project and the Mina Jaya project, and failures in the construction of large export ships in the 1990s, clearly involved failure of the management of the shipyards to appreciate their own technology limitations. It is not always wrong to build ships for export, if the technology is good and the price is competitive, since the new ship market is an international market. But in Indonesia's case, it is said that ships sold for export were almost free ships after taking into account delivery delay cost penalties. Particularly for a country with high interest rates like Indonesia, since the interest rate payments on construction materials push up cost, delivery delay time becomes fatal. Technology is the foundation on which management strategy should be built. If management is carried out disregarding this, there is a high possibility of failure.

10.3.4.4 Methods for Introduction of Technology

Dispatch of JICA specialists was very useful for introducing technological know-how to the shipbuilding industry of Indonesia. However, the JICA specialists could not touch on management matters directly. They were only consultants and advisers. Even in the transfer of component technology the right to choose the technology to be transferred was in the hands of the company where technology was received. Transfer of a certain technology would not be carried out as long as the company did not wish it, even if the JICA specialists were on strong grounds in recommending it. As a general point, it might be said that even if the technology transfer managed to succeed in the short term, for it to succeed in the longer term would require the receiving side to continue to nurture the technology.

These comments bring to attention only the negative side of the JICA technical cooperation. There were also many fruitful points. The technology that JICA specialists taught, for example, in the areas of modernization of advanced equipment and prefabricated block construction, is now becoming established at PT.PAL. Moreover, the survey on which this chapter is based heard, in the private shipyard in Jakarta, that JICA specialist teaching had succeeded in helping to reduce the time needed for repair work. The technical help from JICA is seen in a different light in matters to do with individual support rather than policy.

10.4 CONCLUSION

The lessons that can be learned from the two case studies of technology transfer in shipbuilding can be summarized and discussed under the following headings.

10.4.1 Tacit Knowledge

There are many aspects of shipbuilding that depend on the tacit knowledge of the engineers and workers. Transferring this tacit knowledge is of key importance for successful technology transfer. In the case of Tsuneishi Shipbuilding, more than 100 Japanese engineers and workers were sent to THI for a long period of time. They showed how a task should be done and then gave detailed guidance and support on working methods. In addition, the Japanese engineers returned home in stages, after confirming that the local workers understood the necessary skills. On the other hand, in Indonesia's case, JICA specialists were spread thinly over many sites. One instructor guided many local workers. Because of time limitations, an instructor sometimes had to leave the site before local workers properly understood their work. The tacit knowledge transfer was often not effective.

10.4.2 Construction Strategies

It is important to implement a construction strategy that initially compensates for low levels of technical capability and which allows the knowledge and know-how for construction to accumulate. In the case of THI, only the construction of handy-size bulk carriers was carried out. The design and planning was outsourced to Japan. By means of series construction of the same type of ship, construction knowledge and know-how could be accumulated in the local site. In addition, since Japan was in charge of the design, the performance of the ships was at a level that could withstand global competition. On the other hand, in Indonesia's case, the numbers of same-type ships constructed in one shipyard remained small because many kinds of ships and shipyards were targeted. Therefore, the accumulation of knowledge and know-how could not be carried out effectively. Also, in deciding to target large-sized export ships and designing them in the home country, not enough thought was given to a construction strategy that complemented a low technological ability.

10.4.3 Supply Chain Problems

A reliable supply chain is important. In the case of THI, since the parent company in Japan managed the supply chain, there were no big problems in the delivery of materials and equipment that were needed. On the other hand, in Indonesia's case, where companies in the home country were used, huge problems could occur.

10.4.4 Motivation and Management Problems

Human relations are important. In the case of THI, each of the 15 initial employees was paired with a Japanese engineer and took part in the construction of the shipyard. They were sent to Japan for training for half a year. As a result, they not only could get knowledge and know-how of shipbuilding, but also could get to know the Japanese engineers. However, in Indonesia's case, since the JICA specialists were positioned as advisers and consultants, they were not in a position directly to reach the local workers to teach on the importance of motivation and management problems. They could only work through the approval of local managers.

From the above discussions, the importance of construction and management strategies stands out. It is difficult to transfer all of the tacit knowledge of engineers and workers. Only part can be transferred. In the current global economic environment, all companies are exposed to the same international competition even if their technical capabilities are low. It is important to have construction and management strategies that can withstand international competition while at the same time enabling tacit knowledge to become fully developed. Not only is it difficult to transfer even basic tacit knowledge, but to plan such strategy is quite difficult too. Thus, for successful technology transfer in shipbuilding, it is vital for the transferring side to intervene strongly over some initial period, with a critical mass of activity, on matters including management. As well as transferring basic tacit knowledge, it is essential to build management methods whereby a company can withstand global competition even with low technology.

DISCUSSION QUESTIONS

1. Select a type of industry that relies strongly on tacit knowledge and discuss why the industry has to rely on the tacit knowledge.

2. Investigate the international expansion of the industry selected in the above question, and discuss its differences from and common features with shipbuilding.
3. In this chapter, transfer of tacit knowledge in shipbuilding is described, in which tacit knowledge was transferred as it was. Another way of transfer may also be considered, i.e., changing tacit knowledge to explicit knowledge, then transferring the explicit knowledge. Discuss the merits and demerits of those two ways.

11

Example of Overseas Expansion (Food Machinery)

S company is a global leader in the grain after-harvest industry. It makes all types of processing machinery and plants for that. In 2006, its annual turnover including exports was almost 40 billion yen (0.46 billion USD). In addition, it has set up a number of subsidiary companies overseas. Although S company targets all of the world's three main grains, namely rice, wheat, and maize, its main focus is on activities related to rice.

Although objectives of overseas expansion vary from company to company, for S company the main objective is to expand in the local market overseas. Manufacturing products in Japan and then exporting them is possible. But in S company's technology field, manufacturing in the local area is the typical response to reducing costs. Besides, since it is easier to understand customer needs locally, the concept of local supply for local demand applies.

Regarding where to expand, since expansion in the local market is the main objective, countries with high rice production are first targeted. Such countries generally seek machinery with the most advanced technology and functionality, as they export rice of a higher grade than that consumed in the country itself. Then in which of those countries to expand is determined by average worker wages, inflation, laws, political stability, etc.

Table 11.1 shows for some major rice and wheat producing countries what are their market sizes in terms of tonnages produced and export values of rice and wheat. It also shows in which of them S company has set up subsidiaries. There is further information on the subsidiaries' sales relative to the parent company (the parent company's sales are set at 100), what percent of production went to local consumption, and what percent was imported back to Japan. These data are for the year 2007, as an example.

TABLE 11.1

Background to S Company International Business Expansion

Country	Rice (Paddy) Product Weight (1000 tonnes)	Rice (Milled) Export Value (1000 USD)	Wheat Product Weight (1000 tonnes)	Established Year of Local Company	Sales Ratio	Local Sales Ratio (%)	Exporting Ratio to Japan (%)
World total	657,413	11,187,616	689,946	—	—	—	—
China	187,397	385,281	112,463	1997	7	74	1 ~ 3
India	144,570	2,777,280	78,570	2006	9	—	—
Indonesia	57,157	207	—	—	—	—	—
Bangladesh	43,057	2464	844	—	—	—	—
Vietnam	35,943	1,489,970	—	—	—	—	—
Thailand	32,099	2,913,289	1[a]	1986	7	31	2 ~ 3
Myanmar	31,450	413	158[a]	—	—	—	—
Philippines	16,240	56	—	—	—	—	—
Brazil	11,060	21,662	5886	1999	2	70	0
Japan	10,893	—	882	—	100	—	—
United States	8999	824,846	68,026	1980	11	—	—
UK	—	53,509	17,227	1991	3	—	—

Source: Weight and value data from Food and Agriculture Organization (FAO) Statistics Division for year 2007, accessed March 2010.

[a] FAO estimate.

S company's first South East Asian expansion, in 1986, was to Thailand. Thailand has a high quantity of rice production and is the world's number one exporter in terms of value. As S company's market share increased in Thailand, the company further set up subsidiaries in China, then Brazil, then India. As far as the companies' sales that do not go to local consumption are concerned, they are exported to other countries of the world either directly or through the parent company in Japan. The low import rates to Japan reflect the management policy that increased import rates to Japan would only be pursued once local supply and demand needs were fully satisfied and if excess production with high product quality made it sensible to do that.

In the next sections of this chapter, the following are described: the subsidiary companies' products, how S company quantifies manufacturing effectiveness and costs, some other factors, and a case study.

11.1 THE SUBSIDIARY COMPANIES' PRODUCTS

Grain after-harvest machines include hulling machines, rice inspecting machines, stone separating machines and various other types of sorters, rice polishing machines, and weighing apparatus. It is these that are generally manufactured and sold by the subsidiaries. The typical number of parts in a product, for example the hulling machine, is 700. Figure 11.1 shows a typical after-harvest rice polishing machine.

There are other products that require a high specialist knowledge for their manufacture, such as optical sorters and measuring devices and life science application equipment, that are manufactured and sold by a subsidiary company in America. Up to now these have not been manufactured elsewhere because the subsidiary companies elsewhere have not attained the technical level to manufacture them. These are not considered further in this chapter.

11.2 MANUFACTURING EFFECTIVENESS AND COSTS

This section considers the factors that contribute to changes in costs when manufacturing overseas rather than in Japan. In particular, processing

FIGURE 11.1
Rice polishing machine.

and assembly costs are analyzed. They are the ones that are most different between overseas and Japan. Also, they make up about 30% of the manufacturing costs for S company's products.

A quantitative evaluation of the benefit of processing and assembling overseas rather than in Japan is sought by calculating the relative processing and assembly cost, C_p, that was introduced in Chapter 9. In summary, first the relative manufacturing effectiveness, E_P, in a country is determined from

$$E_p = E_E \times (R_A + R_{NA} \times E_w)$$

where E_E is the facilities effectiveness, R_A and R_{NA} are the automation and nonautomation rates, and E_W is the workers' abilities factor (with E_E and E_W set equal to 1.0 in Japan's case). Then C_p is obtained from

$$C_p = C_T/E_P$$

where C_T is the relative time cost of processing and assembly in a country.

The workers' wages in manufacturing industry is one of the elements that contribute to the time cost. Table 11.2 gives relevant wages country

TABLE 11.2

Manufacturing Industry Average Wage Costs (Male and Female) Considered by S Company

Country	Period	Currency	Sex	2003	2004	2005	2006	2007
Asia								
Japan	Month	Yen	Both	296,500	293,100	292,100	299,600	296,800
			Male	327,800	323,100	323,800	332,300	328,500
			Female	195,800	194,100	190,900	194,600	197,700
India	Month	Rupee	Both	1078.9	1731.8	1234.4	—	—
Korea	Month	1,000 won	Both	2074.0	2280.0	2458.0	2594.8	2772.0
			Male	2369.7	2599.8	2798.6	2931.9	3123.0
			Female	1320.0	1419.7	1556.1	1675.6	1785.0
Thailand	Month	Baht	Both	6432.2	6129.0	6407.4	6941.6	6999.2
			Male	7344.8	—	—	7973.4	—
			Female	5538.8	—	—	5996.6	—
China	Month	Yuan	Both	1041.3	1169.4	1313.1	1497.2	1740.3
Philippines	Day	Peso	Both	237.72	236.65	252.77	274.81	276.52
			Male	249.25	239.77	258.78	277.05	286.62
			Female	221.20	232.08	244.59	271.68	261.84
North America								
United States	Hour	U.S. dollar	Both	15.74	16.15	16.56	16.80	—
Canada	Hour	Canadian dollar	Both	19.76	20.28	20.65	20.76	21.58
South America								
Brazil	Month	Real	Both	901.85[a]	—	—	—	—
			Male	1009.75	—	—	—	—
			Female	618.61	—	—	—	—
Europe								
UK	Hour	UK pound	Both	10.27	10.49	11.16	11.37	—
			Male	10.83	10.95	11.57	11.80	—
			Female	8.62	9.17	9.96	10.05	—

Source: From *World Statistics 2009*, Japanese Management and Coordination Agency Statistics Bureau.

[a] Brazil data for year 2002.

by country considered by S company. They are used in calculating C_p in what follows.

It is also necessary to have data on the workers' abilities factor. Several hundred to several thousand of each type of the company's machines are made in its subsidiaries each year. The typical machine, like the hulling

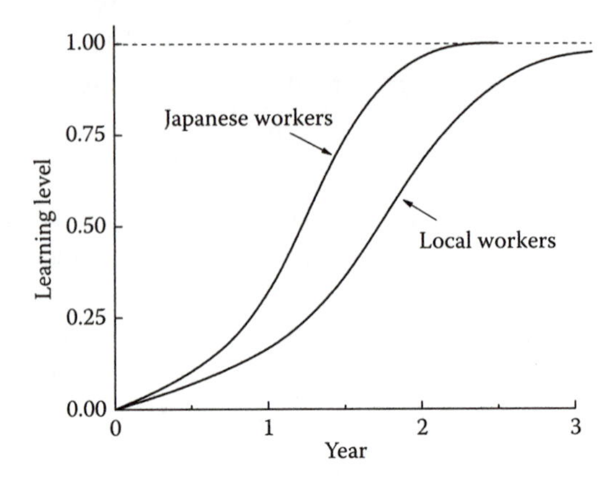

FIGURE 11.2
Concept of technical learning level, with timescale from S company's experience.

machine, has about 700 parts. Subassembly and assembly of a machine are carried out by a worker from start to finish. It is vital that the worker has an understanding of all the parts of the product. In these circumstances, it takes approximately 3 years for a worker's technical skill level to reach the saturation value of 100. The required skill level is much higher and the learning curve longer than, for example, the manufacture of household appliances on a conveyor line, when each job is just the tightening of two or three screws.

Figure 11.2 shows technical skills learning curves for overseas and Japanese workers, based on S company's experiences. The former are found to take approximately 6 months longer to learn than the latter. Furthermore, the final level of the overseas workers is 98 compared to the Japanese 100. The reasons may be the differences in education level and application. Whatever the reason, since more time is required to learn, it leads to a rise in costs.

With these as background, relative manufacturing efficiencies and then processing and assembly costs, at least as far as wages costs are concerned, may be calculated. From Table 11.1, Thailand, China, and Brazil are those where the company expanded. Efficiencies and costs are compared between these and Japan.

Table 11.3 contains the input data on which the comparisons are based. It also contains the results of the comparisons.

First, workers' abilities, E_W, are entered, based on Figure 11.2's values for fully trained workers. Then facilities effectiveness, E_E, are considered to

TABLE 11.3

Relative Effectiveness and Cost Data from S Company's Technology Transfer Considerations

	Country			
Factor	**Japan**	**Thailand**	**China**	**Brazil**
E_W	1.0	0.98	0.98	0.98
E_E	1.0	1.0	1.0	1.0
R_A (processing)	0.6	0.4	0.3	0.4
R_A (assembly)	0.0	0.0	0.0	0.0
R_A (overall)	0.42	0.28	0.21	0.28
R_{NA}	0.58	0.72	0.79	0.72
E_P	1.0	0.99	0.98	0.99
C_T	1.0	0.057	0.059	0.131
C_P[a]	1.0	0.058	0.060	0.132
C_P[b]	1.0	0.068	0.072	0.157

[a] Calculated with $E_W = 0.98$ for Thailand, China, and Brazil.
[b] Calculated with $E_W = 0.77$ for Thailand, China, and Brazil.

be equal to the parent company's across all of the company's subsidiaries, as high-performance machinery is installed and fully used in all of them.

The automation rate, R_A, differs between the processing and assembly stages. First, these are considered separately. Although automatic machines are used in processing, which might be thought to lead to high values of R_A, their use is for high-variety, small-batch processing with much manual materials' supply, loading and unloading of machines, and reprogramming of machines between batches. These all act to reduce the R_A (processing) values in Table 11.3. Assembly is not carried out automatically at all, as it is a high-variety, small-batch activity. Thus, R_A (assembly) is zero.

The ratio of processing to assembly in the total workload is approximately 7:3. The processing and assembly data then lead to an overall value for R_A, as shown as well as to the nonautomation values R_{NA} ($R_{NA} = 1.0 - R_A$).

Taking all the above, with values for the fully trained workers' abilities, E_W, the manufacturing effectiveness, E_P, can finally be calculated. For Thailand, for example, $E_P = 1.0 \times (0.28 + 0.72 \times 0.98) = 0.99$.

The relative time costs, C_T, are obtained from the wages in the manufacturing industry of each country in Table 11.2, relative to Japanese wages.

Finally, C_P is obtained. Continuing the example for the fully trained workers, C_P for Brazil is 0.131/0.99. In this case, it is the lower wage rates in the overseas countries that dominate the calculations.

That remains the case, but not to quite the same extent, if calculations are repeated for manufacturing with workers who are still undergoing training. In that case, the workers' ability index, E_W, has not yet reached 0.98. From Figure 11.2, some average relative ability over a worker's first 3 years can be obtained as the ratio of the areas under the two curves. If that is taken as 1.3:1, E_W reduces to ≈0.77. Then the changed values of C_P in the final row of Table 11.3 are obtained.

11.3 OTHER FACTORS TO CONSIDER

The above calculations show the large advantage of processing and assembly overseas. However, there are disadvantages that can result in larger costs than indicated.

- Product quality: There is a tendency for part quality, other than that relating to basic performance, to be insufficient. For example, a machine door may be poorly fitted or painted, leading to a poor appearance.
- Education costs: Sufficient training facilities are needed, for example computer-based animated manuals, or instructors directly demonstrating how to do things. These take both time and money.
- Communication: When there is a problem with product quality, there may then be problems in communication through an interpreter when investigating the cause and determining how to avoid the problem occurring again.
- Awareness of work targets: For example, at the end of a workday, if there are small jobs still to be done, overseas workers may leave them for the next day, whereas Japanese workers would finish them before leaving.

On the other hand, the learning curve example, Figure 11.2, is a worst case. After a worker has trained for part processing and assembly of a product for the first time, taking about 3 years, if the worker then retrains for a different product, the learning period reduces to about 2 years. Understanding of the details of the technologies of processing, assembling, painting, and general manufacturing deepens as the workers get accustomed to the work. As experience is built, learning periods become even shorter.

11.4 OVERSEAS EXPANSION EXAMPLE: THAILAND

S company's overseas expansion to Thailand is followed in detail, as an example.

Its subsidiary company was set up in August 1986, with a transfer of capital of 42 million Thai baht (126 million Japanese yen) and working funds of 230 million baht (690 million Japanese yen). The factory area was 6289 m^2, with a total area of 11,428 m^2.

Table 11.4 lists the products made by the company and the year that manufacture of each started. The technology transfer from Japan and start of manufacture occurred product by product according to market need.

Figure 11.3 shows sales growth by value and also profits over the years 1987 to 2009. Ten years was needed before the first profit was obtained. However, it was industrial capital, not financial and trading capital, that was invested. It was considered by the company not to be a problem since it was its first experience of overseas expansion. The company's later expansions to China

TABLE 11.4

S Company's Technology and Products Transfer to Thailand, Year by Year

Year	Transferred Technology	Products
1986–1987	Paddy husker	HR10MPc
1988	Rice polisher	KB40
1989	Stone separator	GA50D
1990	Rotary sifter	ST527
1991	Milling machine	IRM30A
1992	Semiautomatic packer	HP15C
1993	Milling machine	RMB10G
1994	Paddy husker	HR10NE
1995	Air screen cleaner	ASC
1996	Vertical-type milling machine	VTA10A
1997	Bag filter	—
1998	Stone separator	SGA10B
	Paddy husker	HR10P(3)-T
1999	Stone separator	SGA5A
2000	Paddy cleaner	PC05D
2001	Milling machine	VBF10A
2003	Rice packer	THP60A, SHP60, SLS20CA, STBS120B, STBS200B
2004	Rice packer	SLS30CA, STBS40B

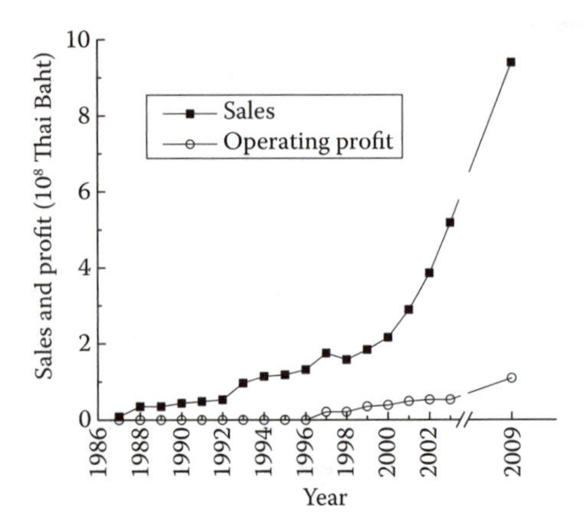

FIGURE 11.3
Sales and profit of S company in Thailand (1 Thai baht ≈ 2.64 JPY ≈ 0.031 USD).

and Brazil showed a profit in a shorter period, as a result of learning from experience and studying the management of other overseas companies.

Figure 11.4 shows the sales value by area, from 1998 onward. In 2004, the sales to Japan are shown as 39.4% of the total and appear to contradict data in Table 11.1. However, most of these were for intermediary trade through Japan. The number of imports for use in Japan was very low.

Figure 11.5 shows the growth in number of employees from 1987 to 2009. It also shows the earnings per employee over that period. When

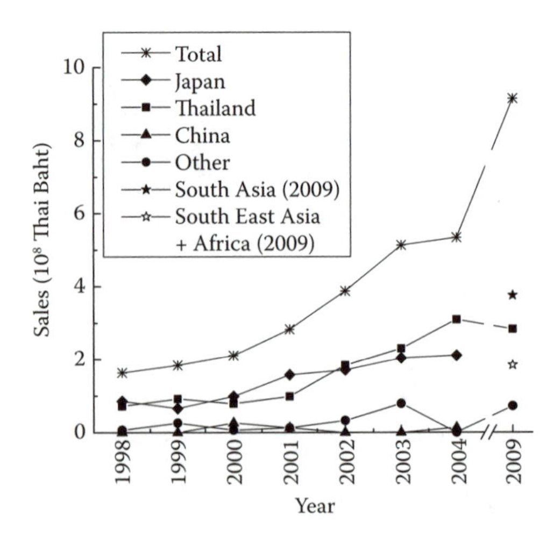

FIGURE 11.4
Regional sales of S company in Thailand (changed regional divisions for 2009).

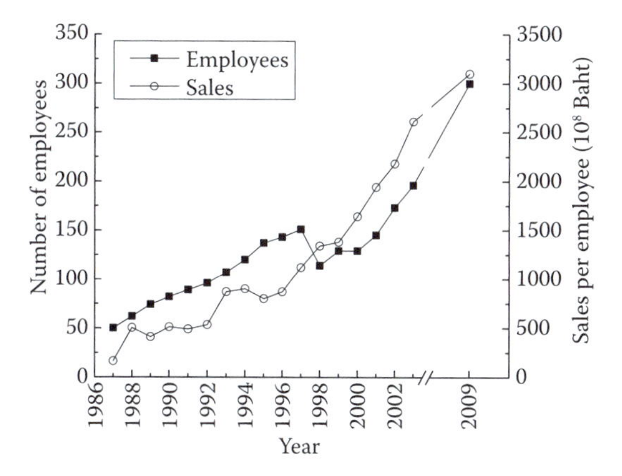

FIGURE 11.5
Number of employees of S company in Thailand and sales per person.

the company was first established, there were only 50 employees, but now there are more than 200, with growth in proportion to sales.

11.5 SUMMARY

This chapter presents the background to a particular company's overseas expansion. The purpose of the company's first venture overseas was to target a country with a high production of rice and to expand on the basis of manufacturing locally for local consumption.

Calculations after the event, using the methods of Chapter 9, show the manufacturing efficiencies and processing and assembly costs for a range of countries (Thailand, China, Brazil) relative to those for Japan. The values obtained depend on the skill levels of the local engineers and technicians and correctly rank the countries in order of their favorability for overseas expansion (Thailand, then China, then Brazil). This supports the validity of the methods suggested in Chapter 9, although it must be said that in this case, unlike in the company A and B examples of Chapter 9, it is the time cost differences between the countries that dominate the calculations.

Finally, annual management (sales, profits, and employee numbers) data on the company's operations in Thailand are presented to indicate, as an example, the long times needed for a successful activity to become established.

DISCUSSION QUESTIONS

1. Explain the reasons why the management of an enterprise may decide to transfer technology. If you have a good knowledge of any particular enterprise, give your answer in the context of that. Further list the countries to which you would consider transfer, with the positive and negative aspects of each.
2. List and explain the items that should be taken into consideration when you judge whether you should transfer technologies or not.
3. After technology has been transferred, how would you judge whether or not it was a successful transfer?

Index

D

E

N

O

Q

R

U

V

W

Y